CROP PRODUCTION TECHNOLOGY

(Rabi Field Crops)

MUKUND JOSHI

Former Associate Professor of Agronomy
College of Agriculture
University of Agricultural Sciences
Bangalore

M.N. THIMMEGOWDA

Professor and Head, AICRP on Agro-Meteorology,
University of Agricultural Sciences,
Bangalore

PHI Learning Private Limited

Delhi-110092
2026

In fond memory of ***Shri Asoke K. Ghosh*** *(October 1942–February 2024), Founder Chairman and Managing Director of PHI Learning, whose vision endlessly inspires.*

The Legacy Continues... .

Published by Pushpita Ghosh, PHI Learning Private Limited, Rimjhim House, 111, Patparganj Industrial Estate, Delhi-110092 and Printed by Syndicate Binders, A-20, Hosiery Complex, Noida, Phase-II Extension, Noida-201305 (N.C.R. Delhi).

₹525.00

CROP PRODUCTION TECHNOLOGY (Rabi Field Crops)
Mukund Joshi and M.N. Thimmegowda

ISBN-978-93-5443-896-7 (Print Book)
ISBN-978-93-5443-930-8 (e-Book)

To

my wife, Mrs. Hema

my daughters, Mrs. Anupama and Mrs. Priya

for their emotional and moral support

while preparing the manuscript of the book

Contents

Part 2
PULSES

Part 3
OIL SEEDS

Part 4
SUGAR CROPS

Part 5
MEDICINAL AND AROMATIC CROPS

Abbreviations

%	:	Per cent
@	:	At the rate
<	:	Less than
>	:	More than
a.i	:	Active ingredient
BC	:	Before Christ
cm	:	Centimetre
CPE	:	Cumulative pan evaporation
DAS	:	Days after sowing
dS/m	:	Desi Siemens per meter
EC	:	Exchangeable cation
FYM	:	Farm yard manure
g	:	Gram
GDD	:	Growing Degree Days
K	:	Potassium
kg/ha	:	Kilogram per hectare
Kg/m^2	:	Kilogram per square meter
l	:	Litre
m	:	Meter
mm	:	Millimetre
M.S.L.	:	Mean sea level
m^2	:	Square meter
ml	:	Millilitre
m.mhos/cm	:	Millimhos per centimetre
MT/ha	:	Metric Tons per hectare
N	:	Nitrogen
^0C	:	Degree Celsius
P	:	Phosphorus
ppm	:	Parts per million
q	:	Quintal
t/ha	:	Tonnes per hectare
Zn	:	Zinc
$ZnSO_4$	:	Zinc sulphate

Foreword

Agriculture sector occupies an important position in the Indian economy from the perspective of nation's food security and providing livelihood to about 52 per cent of working population dependent on it. Of late, Indian agriculture has experienced a paradigm shift from input-based to knowledge-based growth. In this endeavour, it is essential to develop human resources who could harness the rapid development in science and technology.

Indian Council of Agricultural Research, New Delhi and State Agricultural Universities are encouraging teachers to write textbooks to provide comprehensive knowledge and latest developments taking place in the discipline to the students.

Shri Mukund Joshi, former Associate Professor, Department of Agronomy, College of Agriculture, University of Agricultural Sciences, Gandhi Krishi Vignana Kendra, Bangalore, who had been teaching undergraduate and postgraduate courses in Agronomy since 1991, has made an excellent attempt in writing a book titled ***Crop Production Technology (Rabi Field Crops)*** covering nearly 22 Rabi crops grown in the country. I congratulate and complement the efforts put forth by the author.

I hope the book will gain popularity among students as well as among teaching community. Besides, I also wish that the efforts of Shri Mukund Joshi and Dr. Thimmegowda will inspire them, especially the young teachers for writing textbooks, essentially from the view of imparting quality agricultural education.

Dr. H. SHIVANNA
Vice-Chancellor
University of Agricultural Sciences
Bangalore

Preface

The knowledge of field crops is absolutely essential for all agricultural graduates, not only with a view to understand their cultivation practices, but also with the objective to know many academic and scientific details of each crop. National and international issues related to food security, food policies, oilseed availability/production, etc. can be better understood with the detailed knowledge of crops. The knowledge of a crop is complete, only by knowing how it is processed and how the crop is related to the industry or consumer, and what are the quality parameters involved. While studying, the information regarding the nutritional importance and anti-nutritional factors of most of the food crops, is considered essential.

Every undergraduate student in agricultural science has to study the courses related to field crops. In every crop, the new knowledge and new practice are regularly added along with the changing statistics. Hence, latest knowledge about a crop should include all these aspects. More challenging aspect for the teachers offering these courses is how to cover the vast information known in each crop, in restricted time frame of a semester. A comprehensive text comprising detailed information on each crop, primarily emphasising the information to the perceptible level of undergraduate student, can effectively deal with such a challenge. This book is written precisely to deal with this challenge. It is designed and written specifically realising the needs of undergraduate students undergoing such courses and teachers who handle them.

In this book, an attempt is made to cover all important 22 field crops, principally grown in Rabi season described under ICAR's VI Dean Committee approved syllabus covered in two crop courses, regularly taught by the Agronomy teachers. Latest aspects about the spread of the crop, cultivation techniques, varieties, nutrient/water/weed management along with specific climatic/soil requirements and quality aspects of all the crops are covered in a crisp manner, suitable for direct teaching. Special efforts are also made to include quality aspects, anti-nutritional factors, processing and terminologies for specific crops. At the end of each group of crops, review questions along with key answers are also provided, to help the students. The book must be immensely useful to all the undergraduate students and teachers equally. It serves as a study material in the classroom and as a quick reading resource for those who are preparing for competitive examinations.

MUKUND JOSHI
M.N. THIMMEGOWDA

Acknowledgements

I gratefully acknowledge the valuable help rendered by my colleagues, Dr. H.M. Jayadev and Dr. B.V. Jayakumar in scrutinising the manuscript and offering corrective suggestions. I would also like to express my deep gratitude to Dr. S. Bhaskar, Dr. S. Sharanappa of Department of Agronomy and Dr. Ravichandra of Department of Plant Pathology in the University of Agricultural Sciences, for their valuable suggestions.

I would also like to acknowledge the useful statistics/concepts obtained from the websites of Wikipedia and Google. Finally, I put on record the support provided by the University of Agricultural Sciences, Bangalore.

Revision of my first book 'Textbook of Field Crops' to conform that to the latest syllabus approved by the VIth Dean Committee was overdue due to my ill health. It would not have been possible to revise the book to the present shape under a new title, based on the paper name, if Dr. M.N. Thimmegowda, Professor and Head of Agro-Meteorology Department, had not come to my rescue. He is the co-author of this book, I thank him profusely.

MUKUND JOSHI

Part 1
CEREALS

Chapter 1

Wheat (*Triticum aestivum* L.)

1.1 ECONOMIC IMPORTANCE

Wheat contributes 30% towards global food basket, followed by rice (27%) and corn (25%). It is an important cereal food crop supplying energy to major chunk of the population. Nutritionally, it is most important source of energy (carbohydrate content 72%), besides being the only cereal with large protein content (10–14%). It is also rich in niacin and thiamine, carotene and iron (5.3%).

It is also the only cereal having high **gluten**—responsible for stickiness and baking qualities. Gluten is a form of protein. No other cereal is suitable to make bread and other bakery products as wheat flour. Gluten provides structural framework for the familiar spongy and cellular texture for bread and other baked products. Diversified use of wheat products in kitchens, sweet stalls and bakeries has made it a **universal food**, both in rice eating areas and non-rice eating areas; both in wheat growing areas and areas not growing wheat. Number of industries like roller flour mills, ready to eat food industry, biscuits and confectionery and bakery industry depends on wheat. The main consumption of wheat in India is for making rotis and chapattis.

Besides human consumption of grains, wheat straw is a popular fodder for cattle.

How does gluten works?
Gluten is the composite of a ***gliadin*** and **a *glutenin***, which is conjoined with starch in the endosperm of wheat grains. Gluten constitutes about 80% of the protein contained in wheat grains. Gluten is formed when glutenin molecules cross-link to form a sub-microscopic network attached to gliadin, which contributes **viscosity** (thickness) and **extensibility** to the mix. If this dough is leavened with yeast, fermentation produces carbon dioxide bubbles, which, trapped by the gluten network, cause the dough to rise. Baking coagulates the gluten, which, along with starch, stabilises the shape of the final product.

1.2 CLASSIFICATION

Wheat is popularly classified based on their chromosome number and ploidy into three groups:

1. **Diploid wheat:** They have diploid number of chromosomes ($2n = 14$). **Einkorn** *T. monococcum*) is a popular diploid cultivated species.
2. **Tetraploid wheat:** They have tetraploid number of chromosomes ($2n = 28$). **Emmer** (*T. dicoccum*) and **durum** (*T. durum*) wheat are popular cultivated species, which were evolved from wild emmer wheat *T. dicoccoides*. Wild emmer wheat was evolved by natural crossing between two diploid wild grasses—*T. urartu* and *Aegilops speltoides*.
3. **Hexaploid wheat:** They have hexaploid number of chromosomes ($2n = 42$). **Bread wheat** (*T. aestivum*) and **spelt wheat** (*T. spelta*) are cultivated species in this group (Figure 1.1 and Figure 1.2). They were evolved in farmers' fields by natural cross between domesticated emmer/durum wheat with wild diploid grass *Aegilops tauschii*.

Figure 1.1 Bread wheat grains.

Figure 1.2 Spelt wheat grains.

Although, all the above species were domesticated at different times, only few became popular. Further, some species in each group were not domesticated and have remained as wild (ex: *T. macha*, *T. timopheevi*, *T. persicum*, *T. polonicum*, *T. turgidum* and *T. orientale*). Depending on the level of polyploidy, the characters of these wheat differ vastly in terms of gluten content, colour of grain, and hardness. Each of the cultivated species could be further classified into groups of wheat, which essentially differ in terms of grain colour or gluten content or even hardness or total protein content.

Wheat is also classified, based on season of its cultivation, as **spring wheat** (which do not require exposure to chilling temperature during transition from vegetative to reproductive phase), and **winter wheat** (which requires a low temperature of 0–8°C during transition from vegetative to reproductive phase).

Out of the cultivated species of wheat, hexaploid *T. vulgare* (also called ***T. aestivum***) is most widely cultivated, while other hexaploid *T. spelta* is grown on limited scale. Among tetraploid wheat, **emmer wheat** were cultivated in ancient times, while **durum wheat** are second most grown wheat in the world after bread wheat. **Einkorn wheat** area is highly limited.

US classification includes six groups based on the grain colour, hardness and season of cultivation. They are **durum wheat** (hard, translucent and light coloured), hard **red spring** (brownish, hard, high gluten), **hard red winter** (hard, brownish medium protein), **soft red winter** (soft, low protein wheat), **hard white** (light coloured, opaque, chalky and medium in protein) and **soft white** (light coloured, soft, and very low protein).

What is the secret of grain colour in wheat?
Phenolic compounds present in the bran layer of grain are transformed to pigments by browning enzymes imparting brownish or reddish colour to wheat grains. White grains have lower phenolic content and lesser browning enzymes.

1.3 ORIGIN AND HISTORY

Origin of diploid and tetraploid wheat, about 8000 years back, is believed to have occurred in a region between **Tigris** and **Euphrates rivers** (between **Syria and Iraq**). But, domestication of hexalploid wheat is believed to be 7000 years back in regions from **Northern Iran to Northern Afghanistan**. Because, grains of *Triticum aestivum* were reportedly found between Hindukhush and Himalayan mountain ranges about 7000 years back. Reports of wide diversity of *T. spherococcum* in North Western India (Mohenjo-Daro excavations about 5000 years back) indicate that North Western part of India is considered as centre of diversity for the hexaploid wheat.

T. durum was originated in **Central Europe**, as evidenced by diversity. Durum (Latin word durum means **hard**) wheat is known for high proteinless gluten endosperm (Figure 1.3).

Figure 1.3 Durum wheat grains.
(*Courtesy:* Wikipedia on wheat)

T. dicoccum (**emmer wheat**) was domesticated in **South Eastern Iraq** during 7500 BC. Both macaroni and emmer wheat must have entered India some times during 5000 BC, as evidence suggest that they were under cultivation in North Western India, by that time.

Triticale
It is an induced hybrid between wheat (*Triticuma estivum*) and rye (*Secale cercale*), originally bred in Sweden and Scotland. It combines high yielding and high protein characters of wheat with disease/stress tolerance characters of rye. Commercial hybrids were evolved recently. Around 15 million tons of triticale was harvested from Poland, France, Germany and Australia. It can be grown for grains or for fodder. Its protein content is higher than wheat, but glutenin fraction is lesser than wheat.

1.4 AREA AND DISTRIBUTION

Wheat is grown in more than 122 countries of the world. It is grown in regions between Tropic of Cancer and Arctic Circle in Northern hemisphere and South of Tropic of Capricorn in Southern hemisphere. It is predominantly absent in regions near equator (up to 30°). However, it is successfully grown even in this region, in high altitudes. Although, wheat is a highly adaptable crop, successful crop is grown between 30–60°N and 27–40°S up to an altitude of 5000 m M.S.L.

Globally, it is grown on 220 million hectare land with a mean productivity of 3.62 t/ha (see Table 1.1). India, China, Russia and the USA together constitute 44.96% of the global area (Table 1.1), while India leads in area (31.40 million hectare). Germany leads in productivity (7.43 t/ha) and most European countries have higher productivity as compared to tropical/ subtropical countries. European Union constitutes 11.01% area, but contributes 16.70% of global production, due to higher productivity.

Table 1.1 Global Area and Productivity of Wheat (2023)

Country	*Area (million hectare)*	*Productivity (tons/hectare)*	*Country*	*Area (million hectare)*	*Productivity (tons/hectare)*
India	31.40	3.52	China	23.62	5.78
Russia	28.83	3.17	USA	15.08	3.26
Australia	12.96	3.20	Kazakhstan	13.12	0.92
Pakistan	9.03	3.11	Turkey	0.63	1.10
Canada	10.68	2.99	France	4.99	7.20
Germany	2.89	7.43	UK	1.72	8.12
Europe	61.48	4.37	World	220.00	3.62

Global production: 799 million tons
(***Source:*** USDA Foreign Agricultural Service information)

Global supply of wheat is contributed by mainly European Union (133.48 million tons), China (137 million tons), India (111 million tons), Russia (91.5 million tons), USA (49 million tons) and France (35 million tons). Other countries contribute less than 30 million tons (2023 statistics).

In India, it is grown between 11–30°N and from sea level up to 3658 m elevations. It is more successfully grown in all Northern states, while its area in Southern India is negligible due to poor performance. However, India's productivity is lower (3.52 t/ha), as compared to 7.20 t/ha in France and 5.78 t/ha in China and even below the global productivity of 3.62 t/ha.

Indian wheat area is spread in the states of Uttar Pradesh, Madhya Pradesh, Punjab, Haryana, Rajasthan, Bihar, Gujarat and Maharashtra. Its spread and productivity in Andhra Pradesh, Karnataka, and many North Eastern states are poor. Uttar Pradesh alone produces 32% of national production (2021 statistics), followed by Madhya Pradesh (21%) and Punjab (14%). Other states contributing to national wheat production are Rajasthan (9%), Bihar (6%), Gujarat (3%) and Haryana (10%). Remaining states contribute negligible wheat. The share of wheat area in important states: Uttar Pradesh (31%), Punjab (12%), Haryana (8 %), Madhya Pradesh (21%), Rajasthan (8%), Bihar (7.75%) and Gujarat (3%).

1.5 CLIMATIC REQUIREMENT

Temperature requirement

Based on its cold tolerance property, wheat is classified as **winter wheat** and **spring wheat**. Winter wheat requires a specific cold treatment called **vernalisation,** which means exposure to specific range of cold temperatures (0–8°) for a period of 4–6 weeks to initiate flowering. Spring wheat will not require such vernalisation. Best wheat is produced in the regions of **cool** (14–15°C) and **dry climate** during major portion of crop life, followed by warm climate (not more than 25°C) during grain ripening phase. **Warm and moist climates are not suitable** for germination and growth. Excessively high temperature beyond 25°C at ripening phase may reduce grain filling and results in chaffiness.

Growing degree days

Growing degree days required for the completion of crop duration are 1538–1665°C, but till anthesis (70–75% of crop period), the GDD required is 807–901°C. This indicates the requirement of cool temperature and shorter day length.

Latitude

Wheat is a widely adaptable crop and can be grown in **temperate**, **subtropical** and **tropical** climate (at higher altitudes). It can be grown in extreme cold tract even beyond 60°N latitude. It can tolerate severe cold climate, after which a warm climate can induce its growth.

Humidity and rain

Cloudy weather, high humidity may promote **rust disease** and rains after sowing, may promote **seedling blight**. High rainfall during the crop growth may damage the crop. Generally, it prefers rainless climate.

Wheat can be grown in annual rainfall range of 300–1130 mm. In India, wheat is grown in the regions from 100 mm in Rajasthan to the regions of very high rainfall up to 4000 mm in humid Western plains. Around 30% of Indian wheat area is rainfed, in which major rains are received during *kharif* and crop is grown under residual moisture.

Day length

It is a typical **short day plant**, requiring **longer** and **cooler nights** to flower, and in warmer climates, it may fail to flower. But it can be grown from sea level to 5000 m, depending on the season of the growth.

1.6 SOIL REQUIREMENT

Wheat is grown on widely varying soils with good drainage and moisture holding capacity. In India, it is principally grown in vast **Gangetic alluvium belt** and **Indus alluvium** as well as sandy loam soils. But, it is also grown in black soils of Southern and Central India, terrai regions of Himalaya and desert soils of Rajasthan.

Sandy loams are considered more suitable due to their ideal combination of water holding capacity and drainage characters as compared to black soils.

The crop is considered to withstand mild acidity (up to 5.5 pH), as well as mild alkalinity (up to 6 dSm^{-1}). But, the crop is sensitive to water logging.

1.7 BOTANICAL DESCRIPTION OF PLANT

Roots

A set of primary and temporary roots (called **seminal roots**) arise from the base of seed, which helps to establish the young seedling. However, they dies off after permanent roots start developing after 20–25 days. Permanent roots (called **nodal** or **crown roots**) develop just below the soil layer (Figure 1.4), but above seminal roots at 4 leaf stage and are responsible for tillering and leaf development. They are thicker than seminal roots and occur on lower 3–7 nodes. The upper most nodes may be above the soil level and roots from such nodes may not reach the soil. But most of other crown roots are responsible for the growth and development of the crop.

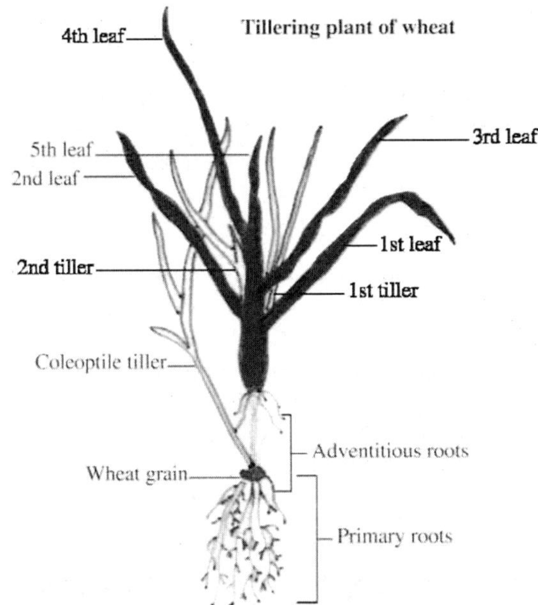

Figure 1.4 Position of tillers and leaves in wheat plant.

Shoots

Shoot consists of round/cylindrical stems with hollow internodes and solid nodes (macaroni wheat internodes may be solid with soft tissues). Main stem is short, but from the lower nodes of main stem, 3–10 secondary shoots grow (called **tillers**), each one having potential to bear the inflorescence (Figure 1.4). Such flowering tillers are called **culm**.

- Invariably basal internodes of each culm are short, while upper internodes are elongated. The intermodal elongation continues till anthesis of inflorescence (Figure 1.5).
- Each tiller may become culm, depending on whether it is fruitful to carry inflorescence. Unfruitful tillers die off to avoid competition among tillers.
- Each tiller bears a sheathed leaf at node, characteristically each sheath clasping the next lower internode. The leaf blade is flattened, lanceolate, dark/light green coloured having parallel venation.

Figure 1.5 Wheat crop.
(*Courtesy:* Wikipedia on wheat)

Leaf

Each leaf blade joins sheath by membranous ligules and sometimes with an auricle. Number of leaves on each tiller decline with the age of its appearance, maximum being on main stem. All productive tillers and main stem end with an ear head/inflorescence (botanically a spike).

Spikelets

They are borne on alternate sides on a zig-zag rachis (Figure 1.6) of spike. Each spikelet may consist of 1–5 florets, each of them is protected by two glumes—**lemma** and **palea**. Awns

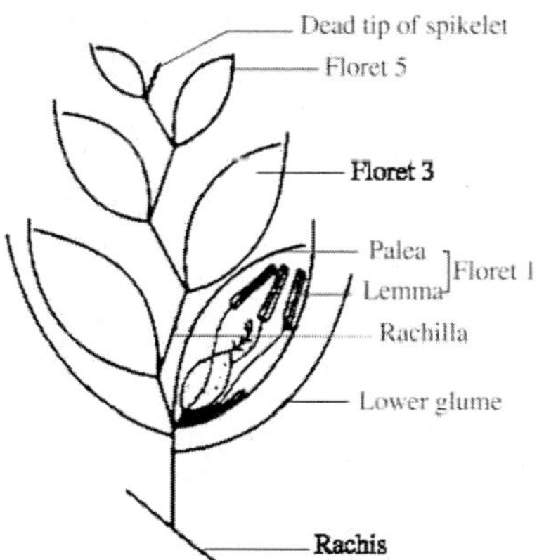

Figure 1.6 Parts of wheat spikelet.

are attached to lemma. Green awns are photo synthetically active and contribute 15% of net photosynthates. Each floret may grow into grain (botanically kernel).

Kernel

Each kernel is a caryopsis oval in shape and grooved (Figure 1.7). It consists of a protein and fat rich **embryo/germ** (2.5% by weight), **bran** (up to 14% by weight), consisting of pericarp, testa and aleurone layer and starch rich **endosperm** (83–87% by weight).

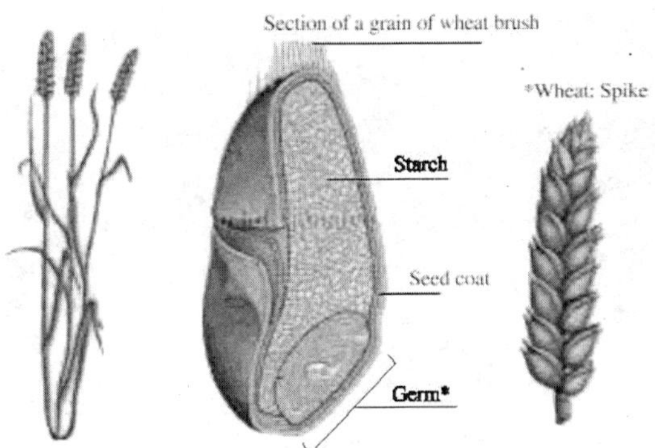

Figure 1.7 Structure of a wheat grain.
(*Courtesy:* Wikipedia on wheat)

1.8 WHEAT GROWING ZONES IN INDIA

Wheat is grown in a number of states of India, which are divided into different zones. Some of the wheat growing zones are presented in Table 1.2.

Table 1.2 Wheat Growing Zones of India

Zones	*Covering Areas/States*	*Description*
North Western Plains	Punjab, Haryana, Delhi, Western Uttar Pradesh, parts of Rajasthan, foot hills of Uttarakhand, Himachal Pradesh and Jammu & Kashmir (40% of area, contributing 55% production in India)	Longest and coolest winter. Crop duration > 145 days. Mid–November sown. More than 95% area is irrigated. Rice-wheat is a dominant system. Rust is a major problem. Major area is under *T. aestivum* and part of area in Punjab **under *T. durum.*** High productivity.
North Eastern Plains	Bihar, Eastern Uttar Pradesh, West Bengal, Odisha, Assam, and other North Eastern states (30% of area, contributing 25% production in India)	Shorter winters. Crop duration 125 days. Late November/December sown, Rice-wheat system is dominant. Medium productivity. Mostly occupied by *T. aestivum.*

(Contd.)

Zones	***Covering Areas/States***	***Description***
Central Zone	Madhya Pradesh, Gujarat, Eastern parts of Rajasthan, Jhansi division of Uttar Pradesh (20% of area, contributing 15% production in India)	Shorter and milder winters. Crop duration 120 days. Both bread and durum wheat are grown. Area known for best quality durum wheat. About 30% area is rainfed (sown in October) and rest is irrigated (sown in mid-November). Low productivity.
Peninsular Zone	Maharashtra, Karnataka, Andhra Pradesh and Tamil Nadu (5% area contributing 2% production in India)	Warmer and extremely short winters. Crop duration 105–110 days. 65% area in Maharashtra and 25% in Karnataka—irrigated (November end sown) and rainfed crop sown in mid-October. Very low productivity.
Northern Hill Zone	Uttarakhand, Himachal Pradesh, Jammu & Kashmir, West Bengal, Arunachal Pradesh and Sikkim (2% area contributing 1% production in India)	Grown up to 3500 m M.S.L. Only bread wheat is grown. Most crop area is under rainfed situations. Low productivity.
Southern Hill Zone	Nilgiri and Palani hills of Tamil Nadu, Kerala	Wheat is grown around the year. First crop: October–April; Second crop: May–September. Area is negligible. Only bread wheat is grown.

North Western plains are most efficient in wheat production, as the growing season extends to more than 145 days (which offers the opportunity to use long duration varieties) besides privileged by 95% of wheat area under irrigation in this zone (Table 1.2). In contrast, Central and Southern zones are known for low productivity, not only because of milder winters, but also, for their large area which is cultivated under rainfed situations. Similarly, Northern hilly zone is also known for poor productivity.

1.9 VARIETIES

Wheat is a self-pollinated crop and no commercial hybrids have yet been released. However, large numbers of varieties are developed for varying situations and against different diseases.

Till the development of semi dwarf wheat (called **norin**) variety by Japanese scientists during 1920, tall low yielding varieties were ruling the world. Movement of such genotypes from Japan during World War II to US later led to develop dwarf wheat varieties in CIMMYT, Mexico. Later, when Dr. Norman Borlaug developed dwarf varieties adaptable to semitropical and tropical regions by using these germplasms, global revolution in wheat production started.

In early 1960s, import of Mexican wheat seeds commenced wheat revolution in India. Based on the early generation material supplied by Dr. Borlaug, selections were made to release varieties called **Kalyansona** and **Sonalika**, during 1970s. In late 1990s, spring wheat varieties were developed. Important wheat varieties developed by AICRP centres for different situations are presented in Table 1.3.

Table 1.3 Zone-wise Recommended Wheat Varieties in India

Zone	*Varieties*	*Adoption*
North-Western Plains	WP 542, PDW 215(d), WH 896(d), PBW 502, DBW 16, UP 2338, Raj 3077 PBW 226, Raj 3765, PBW 373, UP 2425, WR 544, HD 2285 PBW 396, PBW 299, Kundan, PBW 175, C 306 KL1-4, PBW 365, KRL 19	NSIC LSIC NSRC SAL
North-Eastern Plains	HD 2733, HP 1761, K 8804, HD 2733, HD 2824, NW 1012, HP 1731 NW 1014, HUW 234, HW 2045, HP 1633, BDW 14, HP 17, Raj 3777 K 8027, KO 307, HD 2888 HDR 77, K 8962, K 9465 KRL1-4, Raj 3077, PBW 365, KRL 19	NSIC LSIC NSRC LSRC SAL
Central Zone	GW 190, DL 803, HI 8498(d), Rai 1555(d), HI 8381(d), GW 366, MP 4010 HD 2285, GW 173, Swathi, J 405, GW 322, DL 788-2, Raj 3777 C 306, Sujatha, HW 2044, JSW 17, HD 4672(d), HI 1500, HI 1531 KRL1-4, Raj 3077, PBW 365, KRL 19	NSIC LSIC NSRC SAL
Peninsular Zone	DWE 162, DWR 1006(d), HD 2380, AKW 1071, HD 2781, Raj 4037, Raj 4083 HD 2501, DWR 195, H 1977, NIAW 34, NIAW 917, HUW 510, PBW 533 NI 5439, MACS 2846, Bijga yellow, K 9644, AKDW 2997-16 MACS 1967(d) KRL 1-4, Raj 3077, PBW 365, KRL 19 NP 200, DDK 1001, DDK 1009, DDK 1025, DDK 1029	NSIC LSIC NSRC SAL *T. dicoccum*
Northern Hill Zone	HD 240, HD 2380, VL 738, VL 832 HS 207, HS 295, HS 420,VL 616, HS277, VL 289 HPW 42, HS 365, HS 375	ML-NSIC ML-NSRC HL
Southern Hill Zone	HW 971, HUM 318, HW 1085, NUW 2044, NIAW 917, PBW 533, HI 977 NP 200, DDK 1001, DDK 1009	All situations *T. diccoccum*

NSIC = Normal sowing, irrigated condition; LSIC = Late sown, irrigated condition; SAL = Salt resistant; NSRC = Normal sowing, rainfed condition; LSRC = Late sown, rainfed condition; ML = Medium altitude; HL = High altitude.

High yielding wheat varieties have been developed and recommended by Indian Agricultural Research Institute (IARI) for various situations (Table 1.4).

Table 1.4 Wheat Varieties Developed from IARI for Different Zones

Variety	*Recommended for Regions*	*Specific Characters*
HD 2733	North Eastern plain (50 q/ha)	Double dwarf, medium early, stability against all 3 rusts
Kaushambi	North Eastern plain (41 q/ha)	Early maturing, heat tolerant, resistant to all 3 rusts
Purva (HD 2824)	North Eastern plain (46 q/ha)	Adjustable, late sown situations, resistant to all rusts
Urja (HD 2864)	Central zone (42 q/ha)	Suitable to late sown situations, resistant brown rust, foot rot
Pusa vishesh	Delhi (56 q/ha)	For normal sowing situations, resistant to all 3 rusts

(Contd.)

Variety	*Recommended for Regions*	*Specific Characters*
Pusa gold	Delhi (37 q/ha)	For late/extra late situations, resistant to leaf resistance and thermo resistant
HD 2888	North Eastern plain (23 q/ha)	For normal season, resistant to leaf/stem rust
Pusa tripti	Peninsular zone (40 q/ha)	For late sown situations, resistant to leaf/ stem rust
Pusa wheat 111	Central zone (43 q/ha)	For late sowing, high zinc content, resistant to brown/yellow rust
HD 2967	North Western/North Eastern plain (50 q/ha)	For timely sowing, highly adaptable, resistant to leaf blight, resistant to all rusts
Pusa basant	North Eastern plain (35–40 q/ ha)	For irrigated late sown situations, resistant to rusts
Pusa bahar	Peninsular zone (20–30 q/ha)	For irrigated/rainfed late sowings, high gluten
Pusa navgiri	Southern hill zone (52 q/ha)	For irrigated timely sown situations, resistant to rusts, tall and lodging resistant
HI 1563	North Eastern plain (38 q/ha)	Late sowing situations, resistant to all 3 rusts, leaf blight
HD 3043	North Western plain (43 q/ha)	Resistant to all 3 rusts, high gluten

Wheat Hybrids in India and the World

Although, hybridisation program was initiated as early as sixties, CMS approach did not give yield advantage and the program was dropped. By 1995, **chemical hybridising agent (CHA) approach** was adopted by the Directorate of Wheat Research, Karnal. It was found that heterotic advantage (in terms of yield) over best variety in a given agro-climatic zone up to 41% was noticed. In the near future, hybrid wheat may revolutionise the wheat production in India.

(***Source:*** Directorate of Wheat Research, Karnal)

At global level, hybrids have been evolved for advantage in grain, straw yield as well as quality in terms of bread making quality in many countries. Companies like Syngenta, DuPont and Bayer are briskly following up the hybrid program to release the hybrids in wheat. In France, three hybrids, namely, Hyfi, Hyrise and Hypod were released during 2013 for commercial cultivation.

1.10 CROPPING SYSTEMS

Irrigated systems

- **Rice–wheat:** Most widely practiced irrigated system in North Western and North Eastern plains. Wheat is sown by drill or broadcast after rice.
- **Maize–wheat:** Practiced in Jammu & Kashmir, Himachal Pradesh, Uttar Pradesh Rajasthan, Madhya Pradesh under irrigation. Weeds are predominant in wheat. Wheat productivity is low.
- **Cotton–wheat:** Short duration, early sown cotton followed by wheat; Practiced in irrigated parts of Punjab, Haryana, West Rajasthan and parts of Madhya Pradesh. Late seeding of wheat may reduce the yields.

- **Sugarcane + wheat:** Irrigated intercropping system is limited to Uttar Pradesh, Punjab, Haryana, Bihar and parts of Assam, Maharashtra and Karnataka; 3–4 rows of wheat grown simultaneously in between two rows of sugarcane; irrigation and fertilisers to wheat takes care of initial needs of sugarcane. After wheat harvest, sugarcane is fertilised.

Rain fed systems

Principally followed in vertisol regions of Madhya Pradesh, Gujarat, Maharashtra and Karnataka and rainfed regions of Punjab and Haryana. Following are prominent:

Fallow–wheat; pearl millet–wheat; groundnut–wheat; green gram–wheat; soybean–wheat; sesame–wheat and maize–wheat.

Wheat + safflower in vertisols; wheat + barley; wheat + mustard; wheat + gram (sandy loams/alluvial soils); wheat + lentil (in Central and Western India)

Mixed cropping systems

Wheat + mustard; wheat + chick pea + linseed;

1.11 NUTRIENT MANAGEMENT

Wheat requires a well manured soil, not only because it needs nutrients from nature, but also manure application improves water holding capacity, which is essential for a crop that is grown with little water in post monsoon season. It needs to be applied with 8–10 tons of organic manure per hectare.

Tall varieties are not responsive to higher doses of fertilisers and are prone to lodging, leading to poor productivity. Although, triple dwarf wheat can respond up to 200 kg N/ha without lodging, optimum recommendation to grow irrigated wheat is 130 kg N/ha.

Response to higher N application depends on crop duration (which is the function of length of cold temperature), plant population as well as whether crop is irrigated or not. Longer the duration (due to extended cold period), better is the response to added N; higher the population, better is the response to N, especially under irrigation.

It is advantageous to apply half N dose as a basal and half as top dress in most irrigated situations. If the N dose is higher than 120 kg, three splits may be useful. However, under rainfed situations, N dose is reduced to 40–50 kg/ha and entire dose is given at sowing time.

Highest response to added P fertilisers was up to 40 kg/ha (irrigated) and 25 kg/ha (rain-fed). Most P fertilisers are applied as basal for good response. Critical limit for P response ranges between 16 and 50 kg/ha for alluvial and black soil respectively.

Potassium application is useful only when soil K level is low. Response up to 40 kg/ha is reported at 120 kg N/ha. Like P, K fertilisers are given as basal dose.

Application of S and Zn fertilisers depend on the status of soil. Under intensive cultivation, these nutrients become unavailable and they need to be applied at least once in two years.

1.12 WATER MANAGEMENT

Depending on the water holding capacity, 4–6 irrigations are necessary. First irrigation has to be given at **crown root initiation (CRI)**. Other ideal critical stages are late jointing, late tillering, flowering, milking and dough stage.

On an average, water requirement varies from 40–60 cm depending on the crop duration. Generally, a pre-sowing irrigation is necessary for the better establishment of crop. Check basin or border strip method of irrigation are suitable for wheat.

1.13 WEED MANAGEMENT

Critical competition for wheat is initial 30–40 days after sowing. Heavy weed infestation may reduce the tillering, induce competition and reduce growth and yield of the crop. Closer spacing will not allow intercultivation, weed control is rendered more difficult by poor ability of the crop to dominate till 40 DAS. Manual hand weeding after 4 and 6 weeks is efficient, but is a costly operation.

Weeds like *Phalaris* or *Avena* are difficult to be identified in wheat population at initial stages. This makes weed control more difficult. Herbicides may be more economical in such situations. The use of 75% Isoproturon at 1.5 kg/ha is useful as post-emergence (30 DAS) or pre-emergence spray of Pedimethalin @3.3 l/ha (2–3 days after sowing) is useful to control important grass weeds. Broad leaved weeds could be easily controlled by the spray of 2, 4-D at 0.4 kg a.i./ha.

Important Weeds of Wheat			
Phalaris minor	*Avena fatua*	*Lolium rigidum*	*Poa annua*
Polygonum spp	*Chenopodium album*	*Cirsium arvense*	*Convolvulus arvensis*
Fumaria spp	*Melilotus alba*	*Anagallis arvensis*	*Carthamus oxyacantha*

1.14 PESTS AND DISEASES

- **Brown rust:** Severe in North Western/Eastern plains; Small, oval bright orange pustules on leaves or stem or sheath. Spreads fast during warm and humid climates; induces early maturity, produce shriveled grains, ultimately reduce the yields.
- **Yellow rust:** Severe in hilly region, foot hills and North Western region. Oval pustules appear as yellow streaks on leaf/sheath. Develops in early season, and cause damage to seedlings and severely reduce the growth of crop. Spreads faster in cool climates.
- **Black rust:** Reddish brown elongated pustules occur on stem, sheath, leaf or ear heads; spreads faster in warmer humid climates. At maturity, pustules are black and may kill the entire plant or reduce the yields severely. Mostly occurs in Southern, Central or Eastern India.

Control measures for rust

- Use of rust resistant varieties which includes VL-719, HD-2189, UP-2425 and PBW-343.
- Avoid late sowing of late maturing varieties.
- Avoid excess N application, use balanced NPK.
- In severe cases, spray Mancozeb or Zineb 0.2%, when pustules are not fully developed (during the last week of January/first week of February); Second spray after 10 days; Third/fourth spray at 14 days. (All rusts are caused by different species of fungi *Puccinia)*

- **Loose smut:** Caused by fungus *Ustilago*; not noticeable till heading. Typically, produce black powder in place of grains; can spread to mature grains and remain dormant, till it germinates. Control measures include resistant varieties (Raj 2296, HW 657, HS 240, VL 646). Use of disease free seeds, treatment of seeds with Vitavex @2.5 g/kg seed, and solar heat treatment of seeds of infected seeds.
- **Karnal bunt:** Mainly occurs in Himalayan foot hills and terrain regions; it is caused by fungus *Neovossia.* Can be noticed after grains develop. Part of grain is converted into black powder, gives typical foul smell (due to trimethyl amine). Mixture of infected grains even by 3% will result into fishy smell in flour. Control measures include avoiding susceptible varieties, spray fungicide like Tilt @500 ml/ha, use of disease free seeds, avoid monocropping and avoid excessive irrigation during flowering.
- **Alternaria leaf blight:** Lower most leaves show oval, small discoloured lesions, which late develop into dark brown/grey spots. They may spread to sheath, glumes, and awns. Finally, the spots are covered by black spores, which help to spread the disease. Mostly disease spreads faster under warm and humid conditions. Control measures include treatment of seeds with Vitavex @2.5 g/kg seed, or spray Mancozeb or Zineb 0.2%.
- **Powdery mildew:** White powdery growth on the upper surface of leaf, which later turns to grey/brown. Plants stunted, yield is reduced; commonly, occurs in sub mountainous parts of North India and parts of Rajasthan. Control measures include use of resistant varieties (WH-283, HD 2329), burn affected crop (after harvest) or spray Mancozeb and Karathane (4:1 part) @2 g/l.
- **Termites:** Damaged plants completely dry up or give rise to white ears. Can be controlled by mixing with Methyl parathion @25 kg/ha, while ploughing before sowing.
- **Army worms:** Caterpillars feed on tender parts (including ears) during night and hide during day. Controlled by dusting 2% Methyl parathion @25 kg/ha or spraying Carbaryl @3 g/l.
- **Aphids/jassids:** White flies. It suck the leaves and reduce the growth. Controlled by spraying Quinolphos @1–2 ml/l.

1.15 PRODUCTION TECHNIQUES

- **Land preparation:** A well pulverised compact seedbed is essential (achieved by 2–3 ploughing + 2 harrowing). Fall ploughing is useful in fallow–wheat rainfed system. Manure @8–10 t/ha.

- **Sowing:** Use 100 kg/ha seed (row spacing 20–22 cm) or 125 kg/ha (row spacing 15–18 cm); sown at a depth of 5–6 cm using seed drill during second fortnight of November (normal sowing) or end of December (delayed sowing). Treat seeds with Vitavex/ Bavistin/Thiram @2.5 g/kg seed.
- **Fertiliser application:** Apply by drilling 60 –70 kg N, 40–60 kg P_2O_5, 40 kg K_2O per hectare, as basal dose before sowing. Top dress it with another 60–70 kg N/ha in **two splits** (heavy soils) or **three splits** (light soils). Apply 25 kg $ZnSO_4$/ha, if soil is deficient.
- **Irrigation:** Irrigate at crown root initiation (18–24 days), jointing, booting, heading and dough stage by check basin or furrow method using 4–7 cm irrigation.
- **Weed control:** Broad leaved weeds (*Chenopodium, Melolitis, Convolvulus, Carthamus, Cirsium*) and grasses (*Phalaris, Avena, Lolium, Poa* and *Polygonium*) threaten wheat production. First 30–40 days of crop are critical. Use 2, 4-D as pre-emergence spray (against broad leaved weeds) or Isoproturon-against grasses. Prefer to control the weeds by chemical and mechanical methods.
- **Harvesting:** Harvest when grain hardens and straw turns light yellow, dry and brittle. Seed moisture should be 12–13%. Rainfed crop matures earlier than irrigated crop. Harvesting done by hand sickles or harvesters. Yield varies from 3–4 q/ha in peninsular India to 5–5.5 t/ha. Punjab has reported yields up to 7.5 t/ha, and grain straw ratio 1:1.2 (rainfed) or 1:1–2 (irrigated).

Haveli System
In parts of C. India, where vertisols are predominant, rainfed wheat is raised on plots—where water is allowed to accumulate during kharif season. As water holding capacity of such soils is high, successful wheat is grown using stored moisture. This system is called **Haveli System**.

1.16 QUALITY CONSIDERATIONS

Amber coloured, hard, bold and lustrous grains have higher flour recovery.

Major quality traits are 100 grain weight, protein content, gluten content, bread making properties (water absorption quality), and milling quality.

Biscuits, cakes require very soft grain with weak gluten and 8–10% protein; Bread making requires hard grains, with high gluten and higher protein (> 13%); Chapati making atta requires medium hard grains, with medium gluten and 10–12% protein; Noodle making requires soft grains with medium gluten and waxy starch. Sooji making requires hard grains with high protein (> 13) and high gluten content (preferably durum wheat).

1.17 ANTI-NUTRITIONAL FACTORS

Wheat grain can contain phytic acid, trypsin inhibitors, polyphenols as well as tannins as anti-nutritional factors. They can reduce bio-availability of many minerals, zinc, calcium and iron by chelating them. They are more concentrated in aleurone layer than endosperm. However, simple household methods like roasting, cooking, fermentation as well as germination can reduce their level to safe limits.

Gliadin, a gluten protein may cause gastrointestinal disease called **celiac**, in some people allergic to gliadin, when it gets modified by transglutaminase causing inflammatory reaction to small intestine.

REFERENCES

Anonymous (2013), Crop Water Information: Wheat, *FAO WATER*, www.**fao**.org/nr/water/cropinfo_**wheat**.html.

Anonymous (2013), World Agricultural Production, *Foreign Agricultural Service*, USDA, pp. 8–9.

Anonymous (2014), Field Crop Production, *Foreign Agricultural Service*, USDA, http://apps.fas.usda.gov/psdonline/circulars/production.pdf.

Anonymous, 2022, Agricultural statistics at a glance, DAFW, GoI, pp. 29.

FAOSTAT, 2023, https://www.fao.org/faostat/en/#data/QCL.

Herd, Markus, Jeffrey W. White, L.A. Hunt, Simone Grief and Wilhelm Clupeid (2008), *Field Crops Research*, 105th ed., pp. 193–201.

Reddy, S.R. (2006), *Agronomy of Field Crops*, Kalyani, Ludhiana, pp. 128–156.

Singh, Chhidda, Prem Singh and Rabid Singh (2003), *Modern Techniques of Raising Field Crops*, 2nd ed., Oxford and IBH Publishing House, New Delhi, pp. 55–83.

Wikipedia (2014), *Wheat*, http://en.wikipedia.org/wiki/Wheat.

Chapter 2

Barley (*Hordeum vulgare* L.)

2.1 ECONOMIC IMPORTANCE

Barley is an important industrial crop due to its **malting property**, besides being used as a food grain similar to wheat. Malting the barley grains results in beer, whiskey, brandy and other liquors. Hence, barley crop supports many liquor industries.

It is also an important **staple food grain** in Europe and in many cooler semi–arid areas of the world, where wheat is less adapted. It can be used as flour or roasted grains or can be broken as pearl barley (used in soups) or used to prepare weaning food. It is used to prepare 'barley bread' or used as a component in the preparation of 'health foods'.

Barley grains contain 3.9% crude fibre, 1.3% fat, 74% carbohydrate and 12% albuminoids, but do not contain gluten like wheat. It is widely adapted to temperate regions as well as many subtropical regions like wheat. Globally, it is fourth important cereal, although, its area is restricted in India.

Like wheat, its dry straw is used as a popular dry feed for cattles. Grains are also popularly used as poultry feed and other livestock.

2.2 CLASSIFICATION

Ancestral two-row barley, in wild shattering type, belongs to *Hordeum spontaneum*, which later led to the evolution of improved **non-shattering type two-row barley** *Hordeum distyichum*. In two-row barleys, two rows of grains are developed on central rachis (Figure 2.1). They have lower protein and have more fermentable sugars, and hence, desirable for malting. They have long awns.

By mutation, two-row barley, led to more widely cultivated **six-row non-shattering barley** called *Hordeum vulgare*. (They are either awnless or have short awns). They are rich in sugar and usually used for food preparations. Main problem in using barley for food preparations is difficulty in removal of its hull.

Hullless barley was developed later as *Hordeum vulgare* var. *nudum*, which is characterised by easily removable thin hull. This has gained more importance in recent days, as whole grain can be consumed to improve digestible energy of grains.

Figure 2.1 (a) Barley ear heads (b) Two/four-row barley ear heads.

2.3 ORIGIN AND HISTORY

Hordeum vulgare was originated in **Eastern Asia**, comprising parts of **China**, **Tibet** and **Nepal**. Cultivated two-row barley (*Hordeum distyichum*) was originated in Abyssinia (present Ethiopia). Both these were originated from progenitor *Hordeum spontaneum*.

Two-row barley spread to many parts of Europe, Canada, and Russia. But six-row barley spread to China, Nepal, India, Pakistan, Afghanistan, and Tibet and Eastern parts of Russia.

India is familiar to barley since Aryan period, as Sanskrit name **yava** means barley. It was in use by our ancestors since more than 8000 years. As Nepal and Tibet were parts of India then, it was considered as popular food crop, in India.

2.4 AREA AND DISTRIBUTION

It is mainly distributed in Northern hemisphere, namely, Russia (7.6 million hectare), Ukraine (1.4 million hectare), France (1.8 million hectare), Canada (2.69 million hectare), USA (1.03 million hectare), Turkey (3.33 million hectare), while in Argentina (1.58 million hectare) and Australia (3.8 million hectare) of Southern hemisphere it is produced in large scale. Globally, it is grown over 46.25 million hectare and around 146 million tons of barley is produced annually.

Major share of production comes from European countries (83 million tons), Russia (20 million tons), Canada (8.9 million tons), Ukraine (5.5 million tons), Australia (13.5 million tons) and Turkey (7.3 million tons). In India, barley area has dwindled down from 7.8 lakh hectare in 2012 to 6.2 lakh hectare during 2023, mainly due to the replacement of barley by wheat. The productivity of barley in European countries is as high as 2– 8 t/ha, as compared to India (3t/ha).

Uttar Pradesh (31% of area) and Rajasthan (32% of area) are two major states contributing to 63% of total area. Among other states producing barley are Himachal Pradesh (3% of area), Uttarakhand (3% of area). Other states cultivating barley on small scale are Madhya Pradesh and Ladakh. It is mainly a crop of Northern India, almost completely absent from South India.

2.5 SOIL AND CLIMATIC REQUIREMENTS

Barley requires **cool weather** during early stages, but **warm** and **dry weather** during later stages, as the crop approaches maturity. It grows well under **subtropical** and **temperate** regions than tropical areas. It is the only cereal crop that grows successfully and matures at high latitudes (up to 70°N in Finland) and high altitudes (up to 4500 m in Himalayas). As compared to wheat, it is more drought resistant, and matures earlier than wheat and escapes terminal stress conditions. Areas which are moist and warm are not suitable for the barley crop.

All soils suitable to grow wheat are suitable for barley. However, it thrives well on well drained loamy soils with pH range of 6.5–8.0. Barley is more **tolerant to salinity** than acidity. In India, it is being grown on sandy loam soils, alluvial soils, even black soils in plains and hilly regions, mainly on the basis of suitable temperatures and rainfall.

2.6 BOTANICAL DESCRIPTION OF PLANT

Barley belongs to family *Poaceae*, resembling wheat and grows up to a height up to 1 metre.

- **Stem:** Cylindrical, with 5–7 hollow inter nodes separated by solid inter nodes. The intermodal length is shorter at the base and longer in top. The node many be exposed or covered by sheath of leaf immediately above inter node. The tillers vary from 2–5 per plant.
- **Leaves:** Leaves arise from nodes borne alternatively on opposite sides. Leaf consists glabrous sheath clasping the inter node below, broader light green lanceolate rough lamina with small ligules. Auricles are conspicuous.
- **Inflorescence:** It is spike or head borne at the top of stem, having spikelets attached to zigzag rachis. Each spikelet has two glumes and a floret, but spikelets are in a group of three. All three spikelets are fertile in six-row barley, while central spikelet is fertile in two-row barley. Each spikelet has two glumes, ending in an awn. Barley is a self-pollinated crop and on pollination spikelet develops into grain, where glumes cover the grain as husk.
- **Grain (kernel):** It is caryopsis covered by husk with pericarp, endosperm and embryo.
- **Roots:** It has two sets of roots. Shallow set of roots arise near the soil surface and spread laterally to a distance of 15–30 cm right angle to the stem. Second set of roots grow vertically and reach a depth of 0.8–1.5 m.

2.7 VARIETIES

As compared to wheat, rice, and maize, barley is used on a limited scale in India. Its application for malting has retained its demand and very small section of population is using it as a food.

Hence, good barley varieties in India should have malting properties. Different varieties are recommended for different growing zones:

- **North western plains:** Punjab, Haryana, Western Uttar Pradesh, Rajasthan, and parts of Madhya Pradesh.
    - ***For timely sown irrigated areas:*** RD 2552, RD 2503, RD 2035, PL 172, PL 426, Pragathi, Rekha, Alfa-93, Karan 16, DWR 28
    - ***For late sown irrigated area:*** RD 2508, DL 88, Manjula, Preeti
    - ***For rainfed areas:*** RD 2508, PL 419, Harithma, Jawahar, Lakhan
- **North eastern plains:** Eastern Uttar Pradesh, Bihar, Odisha and West Bengal.
    - ***For timely sown irrigated areas:*** RD 2503, RD 2552, Pragathi, Rekha, Ritambhara
    - ***For late sown irrigated area:*** RD 2508, Manjula, Kedar
    - ***For rainfed areas:*** RD 2508, K 603, Harithma, Geetanjali
- **Northern hills:** Hilly regions of Jammu and Kashmir, Himachal Pradesh and Uttarakhand.
    - ***For rainfed area:*** BSH 169, HBL 113, HBL 276, HBL 316, Himani, Dolma, Son VLB 1

Note: All RD series mature in 130 –140 days, yield 40 –55 q/ha and are six-row barley, grown to a height of 100 –105 cm; Karan-16, and Geetanjali are husk less six-row barley; RD 2503, DWR 28, Ritambhara and Alfa 93 are specially bred for malting purposes and yield 40–45 q/ha.

2.8 CROPPING SYSTEMS

Being short duration and drought-resistant crop, and a crop which can withstand early/late sowing situations, barley suits many **sequential systems**. Important sequential systems in practice are: rice–barley; sorghum–barley; pearl millet–barley; guar–barley; groundnut–barley; cowpea–barley; green gram–barley.

Some **mixed cropping systems** popularly grown are barley with chick pea, and barley with mustard or linseed.

2.9 SEEDS AND SOWING

Under normal sowing conditions, a seed rate of 75 kg/ha is sufficient for **plough sole method** or **seed drill method** to ensure row crop at spacing of 22.5 cm. Under late sown conditions or when soil is saline, the seed rate may have to be increased to 100 kg/ha. Apart from these two methods of sowing, cropping the seeds in the open furrows (**Kera method**) is also practiced. Invariably, rainfed sowing needs use of pre-soaked seeds than normal seeds.

Sowing is taken up between October 15 to November 15 and under rainfed conditions, it may be extended up to the end of November. Delay in sowing later than this will drastically reduce the yields.

2.10 NUTRIENT MANAGEMENT

Manure application @5–6 tons/ha 15 days in advance at the time of land preparation is mandatory under irrigated and rainfed conditions for a good crop. In addition, following fertiliser dose is necessary for good crop of barley.

- **Irrigated:** 80:50: 50 kg N, P_2O_5 and K_2O per hectare
- **Rainfed conditions:** 50:30:30 kg N, P_2O_5 and K_2O per hectare
- **Late sowing:** 40:30:20 kg N, P_2O_5 and K_2O per hectare

Half of the dose along with full P_2O_5 and K_2O are applied as basal dose and remaining half N is applied after 30 days at the time of first irrigation.

2.11 WATER MANAGEMENT

Barley is invariably grown as rainfed crop. Wherever it is irrigated, 3–4 irrigations are sufficient for a good crop. Irrigations may be synchronised with active tillering stage (30–35 days), heading stage, flowering stage and grain development stage. In sandy soil, one extra irrigation may be needed. Its water requirement is 35–45 cm.

2.12 WEED MANAGEMENT

It is more advantageous to control the weeds by **cultural operations** than herbicides up to 30 days, after which crop will suppress the weeds. Broad leaved weeds are controlled by the spray of 2, 4-D salt @0.75 kg a.i./ha after 30–40 days–if the weed infestation is severe. In case, the crop is infested with wild oats (*Avena fatua*) and *Phalaris minor*, spray of Isoproturon @1 kga.i./ha or Pendimethalin @3.3l/ha 2–3 days after sowing can help to control them.

2.13 HARVESTING, YIELD AND POST-HARVEST

Due to the shattering problems, the crop should be harvested immediately after ripening. The harvesting is preferably done during morning, by using sickles and threshed–either manually or using machines. The crop may give a yield of 30–35 q/ha of grain and 45 q/ha straw.

2.14 PESTS AND DISEASES

- **Stripe:** Caused by *Helminthosporium gramineum*, it is a widely spread disease, especially seen in regions of low temperature and high humidity (Bihar and Uttar Pradesh). Symptoms include yellow stripes on older leaves (both sheath and lamina), which later turn brown and the tissue dries up causing defoliation. It severely reduces the yield. Controlling the disease include measures, like seed treatment with a mix of Thiram and Bavistin (1:1) @2.5 g/kg seed or soaking seeds in cold water well before sowing during summer for 4–5 hours followed by exposure to bright sunlight for 6–7 hours on a concrete floor.

- **Net blotch:** Caused by *Helminthosporium teres*, its symptoms are seen in the seedling stage. They include brown reticular blotches near the tip of the leaves, which coalesce longitudinally into dark brown strips. Its control measures include either seed treatment with a mix of Thiram and Bavistin (1:1) @2.5 g/kg seed or spray of Mancozeb @0.2% 2–3 times with a gap of 10–15 days.
- **Spot blotch:** Dark brown to black lesions occur on coleoptiles and progress inward. The seedlings may not develop or when developed, the leaves are dark green and short length with excessive tillering. Ears do not develop. Control measures include stricter crop rotation and other measures suggested for net blotch.
- **Rust:** They are more deadly causing very heavy losses. All three rusts of wheat can appear on barley. Brown rust is of minor importance, while yellow rust and black rust are common. The symptoms and control measures are similar to those of wheat.
- **Powdery mildew:** The symptoms include powdery growth over surface of leaves and floral bracts. Later, grey powdery lesions may develop over the entire leaf blade resulting into yellowing, browning and drying of the leaves. The control measures include dusting sulphur @15–20 kg/ha or the spray of Karathane @0.2% at intervals of 15 days.
- **Loose smut:** The symptoms and control measures are similar to those of wheat.
- **Covered smut:** The disease can be noticed at the time of heading, when smutted head emerges out. Hard black mass of smut is developed along with axis. The control measures include seed treatment with Vitavax @2.5 g/kg seed.
- **Aphids:** Insects like aphids may suck the sap and reduce the turgidity of leaves, besides spreading the disease barley yellow dwarf. They can be controlled by the spray of Metasystox or Dimethoate @1 ml/l.

2.15 QUALITY CONSIDERATIONS

The quality of barley is decided based on test weight of barley grains, which depends on the number of growth and yield parameters. The test weight of barley should be between 35 to 45 g/1000 grains. It should also have 12% protein, 70% carbohydrate and 1.3% fat. Barley grains are graded on the basis of widths in three classes:

1. **Thin group** having < 2.2 mm width,
2. **Thick group** having > 2.8, and
3. **Medium group** having width in between.

2.16 PRODUCTION TECHNIQUES

- **Land preparation:** Compared to wheat, barley needs **slightly loose seed bed**. Hence, one deep ploughing, followed by two harrowing is sufficient. The sowing is usually done on flat bed, and hence final levelling is crucial for the establishment of good crop.
- **Sowing:** By **seed drill** or **plough sole method** or **Kera method**, using seed rate of 75 kg/ha or in special situations (like late sowing/saline soils) up to 100 kg/ha at spacing of 22–25 cm between rows. Ideal sowing date is between mid-October to mid-November.

- **Weed control:** By repeated intercultivations up to 30 days. Later, post emergent spray of 2, 4-D @0.75 kg a.i./ha could be sprayed after 30 days of sowing.

REFERENCES

Anonymous (2006), Complete Information on Area and Production of Barley.

Anonymous (2010), World Barley Outlook, http://www.canbar6.usask.ca/files/03_Cuthbert.pd.

Anonymous (2013), World Agricultural Production, *Foreign Agricultural Service*, USDA circular series 12–13, pp. 12. http://www.preservearticles.com/2012020422699/complete–information–on–the–area–and–production–of–barley–in–india.html.

Anonymous, 2022, Agricultural statistics at a glance, DAFW, GoI, p. 29.

FAOSTAT, 2023, https://www.fao.org/faostat/en/#data/QCL.

Reddy, S.R. (2006), *Agronomy of Field Crops*, Kalyani, Ludhiana, pp. 169–174.

Singh, Chhidda, Prem Singh and Rabid Singh (2003), *Modern Techniques of Raising Field Crops*, 2nd ed., Oxford and IBH, New Delhi, pp. 147–163.

MODEL QUESTIONS AND KEY ANSWERS FOR RABI CEREALS

Choose Most Appropriate Answer and Fill in the Blanks

1. Two-row barley is well known for good malting quality, because it contains .... (less sugar/more sugar and more proteins/more sugars and less protein/higher proteins)
2. Wheat grains has .... % of proteins. (5–6/15–20/10–14/3–6)
3. Wheat is well known to contain ...., offering its flour-bread making property. (carotene/gluten/glycol/satarin)
4. Both six-row and two-row barley originated from its progenitor *Hordeum* .... (*sontaneum/santaneum/spontaneum/africana*)
5. Scientific name of six-row barley is *Hordeum* .... (*spontaneum/distyichum/vulgare/indicum*)
6. Wheat supplies .... of the global food production. (28%/21%/30%/19%)
7. .... state has maximum area and production of barley in India. (Punjab/Haryana/Uttar Pradesh/Bihar)
8. Area under barley cultivation has reduced to one-fourth of the area 50 years back in India, as its area is replaced by .... (chick pea/potato/wheat/oat)
9. Popular einkorn cultivated wheat varieties belong to *Triticum* .... (*dicoccum/monococcum/aestivum/vulgare*)
10. Species of cultivated tetraploid wheat are *Triticum durum* and *Triticum* .... (*monococcum/aestivum/dicoccum/vulgare*)
11. Besides *Triticum aestivum*, other cultivated tetraploid species of wheat cultivated on limited scale is *Triticum* .... (*vulgare/spelta/broccoli/levistia*)
12. Emmer wheat was originated in .... (Europe/South East Iran/South East Iraq/China)
13. Triticale is cross between .... and .... (wheat and barley/barley and rye/wheat and rye/wheat and oat)
14. Triticale was essentially developed with the intention of higher yield and .... (improved quality/resistance to drought/stress resistance/higher fodder yield)
15. More than 50% of wheat production takes place in three states, namely, Punjab, Haryana and .... (Rajasthan/Bihar/Madhya Pradesh/Uttar Pradesh)
16. Spring wheat does not require exposure to ...., while it is essential for winter wheat varieties to promote growth. (high temperatures/long days/cold temperatures/high soil moisture).
17. Bread making wheat is expected to have higher .... content. (protein/gluten/mineral/starch)
18. Higher wheat productivity is witnessed in .... zone of India. (North Eastern/Central/North Western/Northern hilly)
19. Exposure to cold temperatures (0–8°C) for 4-weeks to promote the growth is called .... (cold treatment/cold seasoning/vernalisation/vernalation)
20. Wheat requires .... and .... weather for flowering. (cool and dry/cool and wet/warm and dry/warm and wet)

21. Cloudy weather associated with high humidity can accentuate .... disease in wheat. (blight/rust/smut/leaf spot)

22. .... cereal has shattering problem. (Wheat/Barley/ Paddy)

23. Only cereal that successfully grows and matures at high latitudes up to 70°N and high altitudes (up to 4500 m M.S.L) .... (wheat/barley/ paddy)

State TRUE or FALSE by Using 'T' or 'F'

1. In early stages of wheat cultivation, it is difficult to identify weeds like *Phalaris*. (T/F)

2. Barley could be cultivated successfully in any part of India. (T/F)

3. Crown roots are temporary, while seminal roots are permanent in wheat. (T/F)

4. First irrigation is given after 20–21 days of sowing in wheat. (T/F)

5. Gluten in wheat flour is known for its expanding property. (T/F)

6. Gluten forms around 40% of protein contained in wheat. (T/F)

7. Barley is used to prepare breads. (T/F)

Answers

Fill in the blanks

1. more sugars and less proteins
2. 10–14
3. gluten
4. spontaneum
5. distyichurn/vulgare
6. 30%
7. Uttar Pradesh
8. wheat
9. monococcum
10. dicoccum
11. spelta
12. South East Iraq
13. wheat and rye
14. Stress resistance
15. Uttar Pradesh
16. cold temperatures
17. gluten
18. North Western
19. vernalisation
20. warm and dry
21. rust
22. Barley
23. barley

True or False

1. T **2.** F **3.** F **4.** T **5.** T **6.** F **7.** F

Part 2
PULSES

Chapter 3

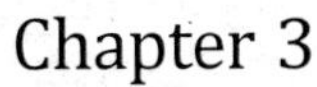

Chickpea (*Cicer arietinum* L.)

3.1 ECONOMIC IMPORTANCE

Chickpea is called **gram** or **Bengal gram** or **Chana** in India and **Garbanzo** in Europe. It is **one of the oldest pulses,** grown in almost all the five continents. It is the most important pulse crop of India, as it is consumed in almost all the states and by all the communities. Its multifarious use in Indian vegetarian food has kept its demand consistently over generations and over locations.

It is mainly a source of **protein** (21%), although, it can supply around **4.5% fat** (more than half of it being polyunsaturated fatty acids), as well as large amounts of **calcium** (202 mg/100 g), **iron** (10 mg/100 g), **niacin** and **Vitamin C** (3 mg/100 g) each. Gram is used as **dry grains** or **unripe green grains** wholly or more popularly as split **dal**. Whole or split gram is either cooked or roasted or converted into flour (called **besan**), used for preparing different sweets/pungent preparations. Even fresh leaves are used as vegetable **sag**.

The husk (for their calcium content) or bits of grains or dry stover are used as popular nutritious **cattle feed** (especially for milch cattle). Chickpea is also considered to have medicinal values for **blood purification** or regular menstruation. It has **low glycemic index**, hence suitable for diabetic patients. In addition, growing Bengal gram improves the **N status of soil** (due to biological nitrogen fixation) and increase the organic carbon status of soils. Therefore, cropping systems involving Bengal gram is more sustainable.

3.2 ORIGIN AND HISTORY

It originated in **South Eastern Turkey**, as 7500 old remains of gram were found in Turkey. This was corroborated by the observation that its progenitor ***Cicer reticulatum*** is found growing only in South Eastern Turkey. Chickpea has **no wild form**.

It must have spread to Europe from Turkey and later spread to Asian countries like India, Pakistan, sometimes during 2000 BC. Its entry into Africa, USA, and Australia is more recent. It was not grown in peninsular India till 500 BC, after which it must have been introduced from Northern India.

In many scholarly Sanskrit literature, as well as popular Sangam literature of Tamil, mentions about chickpea, its cultivation and use of Bengal gram are made frequently.

3.3 AREA AND DISTRIBUTION

Globally, it is grown on about 14.09 million hectare (2023 data). About **75% of the world chickpea** area is in South Asia. India is, by far, **largest chickpea producing** country. Other important chickpea producing countries are Pakistan, Turkey, Mexico, Canada and Australia. India shares 74% of the global chickpea area, with 74% of the global production. Despite being a crop of the subtropical regions, advances in plant breeding have enabled chickpea cultivation to gradually spread to the tropical regions of Africa, North America and Oceania. In Africa, it is grown mainly in Tanzania, Kenya, and Sudan to utilise fallow lands.

Major **exporters** of chickea include Turkey, Australia, Mexico, Canada, Syria and USA—exporting annually around 950 thousand tons. **Importing countries** mainly include India, Pakistan, Sri Lanka, Bangladesh, Spain, Algeria, Tunisia, Jordan, Italy and UK. Asian countries import brown, small seeded chana, while other countries import white Kabuli chana. Global chickpea production has increased over the past 15 years from **10.9** million metric tons to **16.51 million metric tons**.

In India, it occupies around **30%** of Indian pulse area, contributing **48%** of total Indian pulse production. Indian production is around **to 12.26 million metric tons**, with an increase in area from **8.65 to 10.47 million hectare** during last 15 years. It is mainly grown in the states of Maharashtra, Madhya Pradesh, Uttar Pradesh, Rajasthan, Karnataka, Gujarat, Andhra Pradesh and, which jointly account 95% of the Indian area (Table 3.1). However, in recent years, many Northern states have lost chickpea area (as per AICRP on Chickpea) and it was compensated by increasing area in Andhra Pradesh, Karnataka and Maharashtra. These states have also recorded steep increase in their productivity.

Table 3.1 Area of Chickpea in Different States (2022)

States	*Area* (m. hectare)	*States*	*Area* (m. hectare)
Maharashtra	2.83	Madhya Pradesh	2.03
Rajasthan	2.25	Uttar Pradesh	0.62
Andhra Pradesh	0.45	Karnataka	0.71
Gujarat	1.10	All other states	0.33

3.4 CLASSIFICATION

- **Brown grain (Microsperma):** (*Cicer arietinum* L.): ($2n = 14$), seed coat is rough and yellow to dark brown; smaller and angular seeds; cotyledon colour–cream, black, brown, yellow or green. Plants are smaller bushy with small leaflets and good branching with purplish flowers. Grown in India, Bangladesh, Pakistan, Ethiopia, Mexico and Iran.
- **White grain (Macrosperma):** (*Cicer kabulium* L.): ($2n = 16$), seed coat is smooth and creamy white; seeds are round, bold and attractive; cotyledon colour–creamy. Plants are taller, erect growing with large leaflets, with white flowers and good branching. It has more sugars; grown in Southern Europe, Latin America (Peru and Chile), Afghanistan and parts of Pakistan/India.

3.5 SOIL REQUIREMENT

It is grown on wide range of soils, varying from heavy clay rich **vertisols** to alluvial, clay loam or even light **sandy loam** soils. But, **well drained** medium soils (**clay loam**) are most suitable to grow gram. Water holding capacity of soil becomes decisive for the success of rainfed crop (most of the gram area is rainfed), because it is grown in post-monsoon rainless period. Hence, **well drained clay rich** soils of high water holding capacity may be used in rainfed areas.

The crop requires neutral soils, the crop cannot withstand acidity, but tolerate mild alkalinity (up to pH of 8.0).

Marginal soil of poor fertility results in poor yield. Similarly, soils with shallow depth are generally not suitable to grow gram, having deep root system.

3.6 CLIMATIC REQUIREMENT

Essentially, it is a **subtropical crop**, requiring cold temperatures for its growth and flowering. However, temperatures less than 5°C or more than 30°C can damage the crop.

Frost during flowering can inhibit fertilisation and fruit formation or kill the seeds inside the pod. Ideal temperatures required for its growth include 24°C–30°C. The flowering requires warmer climates, while in the initial stages, temperatures lesser than 25°C (up to 15°C) do not interfere in the growth process.

Bengal gram is **very sensitive** to **excess moisture** (hence, suitable to be grown in rainless *rabi* season), **high humidity** and **cloudy weather**. It is a **long day plant** requiring a day length of 12–16 hours during flowering. Its growing degree days vary from 1400–1600 degree days (above base temperature of 10°C).

3.7 BOTANICAL DESCRIPTION OF PLANT

It is an herbaceous annual plant of *Fabaceae* family, growing to a height of 20–60 cm.

- **Germination:** Epigeal germination with epicotyl, growing erectly to produce branches and hypocotyl growing into tap roots. Cotyledon develops into first set of leaves called cotyledon leaves.
- **Roots:** It has strong tap root system, the main root growing up to 1.5–2.0 m (Figure 3.1) with 3–4 sets of secondary and tertiary branches, each with starch rich parenchymatous tissues. They get hardened at maturity due to deposition of cork. The secondary/tertiary roots develop nodules, on infection by *Rhizobium*.
- **Stem:** Erect, branched, viscous, hairy, herbaceous, green and solid and succulent when young, but lignified at later stages. Primary branches (1–8) are woody, laterally growing. But secondary branches (2–12) are less woody, but determine the total leaf area of plant.
- **Leaves:** Petiolate, compound, imparipinnate with grooved 3–5 long rachis (Figure 3.2). 10–15 leaflets on each compound leaf are borne on secondary and tertiary branches. Leaflets are opposite/alternate with a terminal leaflet. Each leaflet has ovate/elliptical with serrated margin and rounded top. All the leaflets are pubescent.

Figure 3.1 Rooting habits of chickpea. **Figure 3.2 Imparipinnate compound leaf of chickpea.**

- **Inflorescence:** The solitary flowers are borne in an **axillary raceme**. Sometimes, there are 2 or 3 flowers on the same node. Such flowers possess both a peduncle and a pedicel. The peduncle is 6–30 mm in length. The bracts are 1–5 mm in length. Flowers are small, white or reddish with blue, violet or pink veins and are bisexual having papilionaceous corolla.
- **Pollination:** Cross pollination due to difference in the height of stamen and stigma.
- **Pod development:** Pod develops after 5–6 days of fertilisation. Pods are inflated roundish (1.5–3 cm long, 0.8–1.5 cm wide) **ending with hard mucro**. Pod number: 30–150, depending on the genotype.
- **Seeds:** Spherical in shape, **wrinkled with a beak**. They vary largely in size and colour. Seed colour varies from white, yellowish orange and dark brown. Seed coat may be smooth or wrinkled. Cotyledons are thick and yellowish in colour.

3.8 SEASONS OF CULTIVATION

In peninsular India, where winters are warmer, it is grown in *rabi* season and its duration is as less as 90–110 days. But in northern subtropical belt, the duration of *rabi* crop can extend up to 160–170 days, because of shorter day lengths during initial period.

In Mediterranean and temperate regions, it is generally grown during spring, to avoid extreme temperature.

3.9 VARIETIES

Widely varying climates and soils used to grow chickpea needed different varieties for different growing zones. Some important varieties released in different states are presented in Table 3.2.

Table 3.2 List of Recommended Chickpea Varieties

States	*Recommended Varieties*
Andhra Pradesh	Vicar, Kristi, Sweat, ICCV 11
Bihar	KPG 59, Pusa 372, KWR 108, Pusa Kabuli, GCP 105
Delhi	Andhra, Pusa 1088, Pusa 1103, Pusa 372, PDG 186, RSG 888
Gujarat	RSG 888, Vijay, Pusa 391, Visual
Haryana	Andhra, Haryana Kabuli 2, Haryana channa 5, Pusa 362, PDG 186, GPF 2
Karnataka	Vishal, Annigeri-1, KAK-2
Madhya Pradesh	JGK Kabuli, JG 130, JG 322, JG 130, Jawahar gram 1, Vijay, Pusa 391, BGD 72
Maharashtra	RSG 888, Pusa 391, Sweat, Vijay, Viral Kabuli, Jawahar gram 1, Gulch 1
Odisha	GCP 105, Bharathi, L 551, Vihar, Phuli G 9311 Kabuli
Punjab	Aadhar, RSG 888, KGP 50, L 551, GPF 2, Pusa 362, PGD 3, DCP 92–3
Rajasthan	Aadhar, RSG 888, Asha, Pratap channa, Arpitha, PBG 1, RSG 44, Pusa 362
Tamil Nadu	Co 3, Co 4, Vihar
Uttar Pradesh	Aadhar, RSG 888, KPG 59, PBG1, Pusa 372, Pusa 362, KWR 108, Pusa Kabuli
Uttarakhand	Aadhar, ICPL 66039, PBG 1
West Bengal	GCP 105, RSG 888, KPG 59, Bharati, Pusa 372, Pusa 362, KWR 108

3.10 NUTRIENT MANAGEMENT

Nutrient management of chickpea is more complicated and not easily understood, due to complexities in relation to its N-fixing abilities as well as complications caused by varying soil moisture—which obviously decides the nutrient availability. A grain yield of 1370 kg removes 82 kg N, 18 kg P_2O_5 and 76 kg K_2O per hectare. But, application of higher doses of nitrogen has not resulted in yield increase in India and other countries.

Most of the nitrogen requirement is met by the biological nitrogen fixation—provided organic carbon status and soil moisture status are congenial. If not, the crop may respond to the application of nitrogen up to 20 kg/ha in initial stages of crop.

Response to added phosphorus is generally consistent and predominant in most P poor soils. Application up to 40–60 kg P_2O_5/ha has found good response. But application of potassium has produced inconsistent results over years and over locations. In light soils, high potassium application up to 40 kg/ha is recommended. But in many clay rich soils, the response is inconsistent and not recommended. Application of zinc sulphate up to 25 kg/ha in intensively grown crop is useful.

As most of the area is not irrigated and grown in post rainy period, the success of utilising nutrients solely depends on **soil moisture holding capacity of soils**. Clay rich soils, holding more moisture for longer period may respond well to the applied fertilisers.

3.11 WATER MANAGEMENT

As most of the chickpea is grown without irrigation, issues on water management are meager—except that adequate measures to conserve rain water are taken up. However, in case, soil moisture is limited at sowing time, a pre-sowing irrigation may be given. In case of deficit soil moisture, pre-flowering and pod development stages will be useful, especially in light soils.

But, irrigation up to one month of crop establishment or at the time of flowering is not desirable—as it may promote excessive vegetative growth at the cost of reproductive growth.

3.12 WEED MANAGEMENT

Weed management in chickpea is mostly achieved by **intercultivation**, with a hoe/harrow, at an early stage, at 25–30 DAS followed by **hand weeding**. In case of severe infestation, a second intercultivation may be necessary after 50 DAS. Wherever, hand weeding is more costly, it may be useful to take up **pre-emergence** spray of **Pendimethalin** (@1 kg a.i./ha) or **Metalochlor** (@1.5 kg a.i./ha) or **Fluchloralin** (@1 kg a.i./ha)coupled with intercultivation.

3.13 HARVEST AND POST-HARVEST

The time to harvest the crop is indicated by the leaf colour turning to brown and they wither away. The crop is cut at the base or the whole plant is pulled. Then, they are **sun dried** for 5–6 days and threshed by beating or trampling under cattle feet.

Chickpea has a great yield potential (up to 25 q/ha) under irrigation. But in most rainfed situations, its yield seldom exceeds 10–15 q/ha.

3.14 PESTS AND DISEASES

- **Pod borer:** It is a serious pest of Bengal gram; typically, it feeds on developing seeds by boring the pod and entering half way into the pod. It can be highly devastating (yield reduction up to 60–70%). It can be controlled by spraying Monocrotophos (@1–1.5 ml/l).
- **Cut worm:** It cuts the young plants at the base and cause serious damage at the establishment stage. It can be controlled by using Carbofuran (@10–15 kg/ha to soil).
- **Wilt:** It is caused by Fusarium orthocerus. It can attack at any stage of the crop. Sudden wilting of plants, coupled with yellowing, associated with blackening/rotting of root are symptoms. It can be controlled by the seed treatment with Thiram (@2.5 g/kg seed) or deep sowing (> 10 cm).
- **Sclerotia blight:** It is caused by *Sclerotinia sclerotiorum* and damages all parts of the plant. The affected plants turn yellow, then brown and finally dry. The disease is characterised by black spots in affected regions, which girdle and cut off the food supply. It can be controlled by the application of mixture of Brassicol and Captan (@10 kg/ha), besides precautionary measures to avoid mono cropping/use of resistant varieties.

3.15 QUALITY AND NUTRITIONAL CONSIDERATIONS

Chickpea has all the essential amino acids, except sulphur containing amino acids. Fat in chickpea will not pose health problems, because its fat is rich in unsaturated fatty acids like linoleic and oleic acid. Calcium, magnesium, phosphorus and potassium are also present in chickpea seeds. Chickpea is also a good source of important vitamins such as riboflavin, niacin, thiamin, folate and the vitamin A precursor, β-carotene. It has beneficial effects in mitigating important human diseases like cardiovascular disease, Type 2 diabetes, digestive diseases and few types of cancers.

3.16 ANTI-NUTRITIONAL FACTORS

Like other pulses, chickpea seeds also contain anti-nutritional factors, such as trypsin inhibitors, haemagglutinin, tannins, saponins and phytic acids. Phytic acid could be reduced by germination, and others can be reduced or eliminated by different cooking techniques.

3.17 PRODUCTION TECHNIQUES

- **Land preparation:** One ploughing followed by 2–3 harrowing and planking. Application of 5–6 t/ha of farmyard manure is necessary before second harrowing.
- **Seeds and sowing:** First week (peninsular India) or third week (North India) of October is ideal. Sowing by drill or plough sole method, at 30 cm row spacing, along with fertilisers supplying 20 kg N; 40 kg P_2O_5: and 20 kg K_2O per hectare (entire as basal) using 60–70 kg seeds/ha. In Kabuli channa seed rate is higher (80 kg/ha) and inter row spacing is 45 cm.
- **Irrigation:** Whenever grown under irrigation, crop may be irrigated at branching, flowering, pod formation stages and pod filling stages.
- **Weed management:** Inter cultivated 2–3 times before 50 days coupled with one hand weeding. In case, the infestation is severe, pre-emergence spray of Fluchloralin or Pendimethalin (@1 kg a.i./ha) coupled with intercultivation.
- **Plant protection:** Pod borer-Monocrotophos (@1.5 ml/l); Cut worm–Carbofuran (@15 kg/ha); Fusarium wilt-seed treatment with Thiram (@2.5 g/kg seed); Bligh-soil application of mix of Captan and Brassicol (@15 kg/ha).

REFERENCES

Agriwatch (2010), http://www.agriwatch.com/getreportpath.php?db_product_id=200017&type=f&file=Pulses–Weekly–Research–Report–20111214.pdf.

Anonymous (2010), Growing Chickpea in the Northern Great Plains, http://www.fertilizer.org/ifa/content/download/8942/133652/version/1/file/chickpea.pdf.

Anonymous (2011), Chickpea, http://www.icrisat.org/what–we–do/learning–opportunities/lsu–pdfs/sds.16.pdf.

Anonymous (2012), Chickpea, CGIAR, http://www.cgiar.org/our–research/crop factsheets/chickpea/. Anonymous (2012), http://oar.icrisat.org/6107/2/BJN_Chickpea_Paper_Gaur_etal._2012_ Review. pdfttp://www.ag.ndsu.edu/pubs/plantsci/crops/a1236.pdf.

Anonymous (2014), Chickpea, Wikipedia, http://en.wikipedia.org/wiki/Chickpea.

Anonymous, 2022, Agricultural statistics at a glance, DAFW, GoI, p. 29.

El-Adawy, T.A. (2002), "Nutritional Composition and Anti-Nutritional Factors of Chickpea Undergoing Different Cooking Methods and Germination", *Plant and food for human nutrition,* Vol. 57, 1st ed., pp. 83–97.

FAOSTAT, 2023, https://www.fao.org/faostat/en/#data/QCL.

Chapter 4

Lentil (*Lens culinaris* M.)

4.1 ECONOMIC IMPORTANCE

Locally, lentil is called **masoor dal** or **kesari dal** due to deep red orange cotyledons. It is an important cold loving pulse, grown and consumed in many parts of the world including Northern India, Latin America and countries around Mediterranean region. It is mainly consumed as split dal, although, some consumption is reported as whole pulse.

Lentil grains contain 25% protein and 60% carbohydrates and is low in fat. Besides, they are rich in phosphates, potassium, calcium and carotene. Hence, lentils are considered as nutritious pulse. Use of lentil to reduce blood cholesterol, is recommended, because of saponin contained in their seed (0.3–0.4%). It is also useful for diabetic patients due to its high fibre content.

Its use as cattle feed is limited, although, dry haulms and chaffy pods are fed to cattle. Lentil is popularly used as a **cover crop** to conserve the top soil like cowpea. It is also a popular dry land *rabi* pulse, which can be grown in scanty rainfall conditions.

4.2 CLASSIFICATION

In earlier literature, it was called ***Lens esculenta ME***, but now renamed as ***Lens culinaris***. Lentil is classified in two groups:

1. **Bold seeded Macrosperma** (**Masur** or **malka masur):** Characterised by large flat pods with bold flat seeds (around 7 mm diameter), principally found in different countries of Mediterranean region, Africa and Central Asia.
2. **Small seeded Microsperma** (**Masuri**)**:** Characterised by small plumpy pods with small lens-shaped seeds (3–6 mm diameter), principally grown in India, Pakistan and South/West Asia.

4.3 ORIGIN AND HISTORY

Macrosperma lentil was originated in **Mediterranean region** (Asia Minor and Egypt), while microsperma lentil was originated in the **valleys of Hindu kush mountain** between Pakistan and Afghanistan.

Bolder lentil (macrosperma) was cultivated in Egypt and Greece, since pre-historic times. It must have later spread to the different parts of Europe and West Asian countries. Even, smaller lentil was also grown in Pakistan and Northern part of India, since pre-historic period. It must have later spread to the other parts of Asia including China.

4.4 AREA AND DISTRIBUTION

Although, it is considered as fourth important globally grown pulse crop, its distribution is restricted to specific pockets like Northern India, Pakistan, China as well as Egypt, Turkey, Syria, Spain, USA, Morocco, Ethiopia and Canada.

Around 1.6 million hectare out of the global area of 5.7 million hectare, is in India, while remaining area is spread to other countries. Out of the global production of 7 million tons, Indian production accounts for 21%, while Canada produces around 23%. But, highest productivity is found in Egypt 2.4 t/ha, followed by the USA (1.2 t/ha). In India, its productivity is as low as 952 kg/ha. In India, mainly in the states of Uttar Pradesh, Bihar, Madhya Pradesh and West Bengal (they account for 95% of the production and area).

It is predominantly absent from Southern and Western Indian states.

4.5 SOIL AND CLIMATIC REQUIREMENTS

It is a **cool season winter crop** requiring cool climate during its pre-flowering phase and relatively warmer climate during pod development and maturity stages. However, very low temperature (< 10°C) may reduce the rate of growth or very high temperatures (> 30°C) during maturity period may result in chaffiness. But, germination requires cold temperatures between 5–15°C. Ideal temperatures for successful crop growth vary between 18–30°C. But, it can withstand frost and can be successfully grown under low rainfall situations (300–400 mm).

It remains unaffected by rain at any stage of crop growth and it can be grown even without rains (using stored soil moisture). It can be grown up to 3500 m M.S.L. It is a **long day plant** requiring at least 12–14 hours for flowering.

Lentil can be grown in most soils ranging from sandy to clayey–with a condition that they must be well drained. It can tolerate both acidity and alkalinity (pH range 4.5 to 8.2) and can be successfully grown in any marginal land.

4.6 BOTANICAL DESCRIPTION OF PLANT (MICROSPERMA LENTIL)

It belongs to family *Fabacea*. It is a small herbaceous annual erect growing plant (Figure 4.1), which seldom grows beyond a height of 60 cm with 4–6 branches. But, it has well developed tap root system with lateral branches growing in all directions. Stem is quadrangular in cross section but weak; bearing small compound pinnate leaves with 5–7 pairs of sessile leaflets (Figure 4.2). The terminal leaflet of each compound leaf is converted into tendril. The flowers are borne on raceme (with 2–4 flowers) invariably located at leaf axils. The flowers are tiny, white tinged with blue or violet or pink colour, with self-pollination a rule. Anthesis and pollination occur before flower opens.

Pods are short and **two seeded**, with light brown coloured lens-shaped grains.

Figure 4.1 Lentil crop in flowering stage.

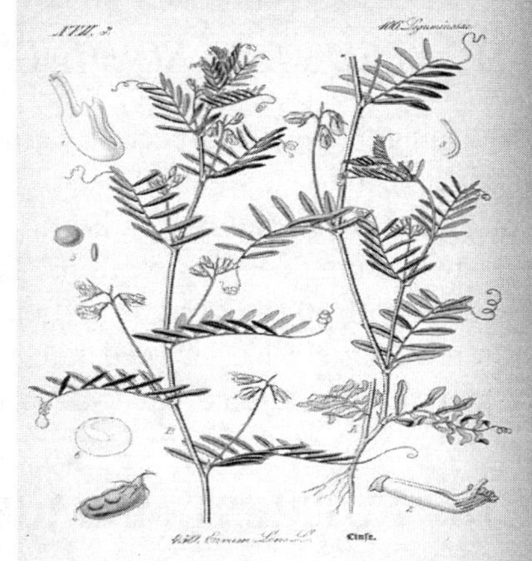

Figure 4.2 Pinnate leaves of lentil.

4.7 VARIETIES

Important lentil varieties grown in different states are presented in Table 4.1

Table 4.1 List of Important Varieties of Lentils Grown in Different States

States	*Varieties*
Bihar	HUL 57, KLS 218, DPL 15, Pusa vaibhav, Sapna, K 75, Pant L 639/406
West Bengal	HUL 57, KLS 218, DPL 15, Pusa vaibhav, Pant L 4, L 4076, Sapna, K 75, Pant L 639/406, Subrata, Ranjan, Asha
Madhya Pradesh	IPL 81, JL 3, Subrata, L 4076, Malik, JL 1
Uttar Pradesh	DPL 15, Pusa vaibhav, IPL 81, JL 3, Subrata, Lens 4076, Sapna, JLS 1, Pant L 4, K 75, Pant L 406, Narendra masoor, HUL 57, DPL 62
Uttarakhand	PL5, Mansoor 103, Vimasoor 4
Jharkhand	HUL 57, KLS 218, DLP 15

4.8 CROPPING SYSTEMS

Lentil is popularly grown as **relay crop**, before harvesting paddy crop in **utera** system of cultivation in the states of West Bengal, Odisha and Bihar. In such system, seeds of lentil are broadcasted in between paddy lines to be harvested. Relay system becomes necessary to use low temperatures available at that time to ensure good germination of lentil.

Besides, the crop is also grown in sequence with maize, cotton, sorghum, pearl millet or even groundnut (all grown in *kharif* season) as sole crop. It is also grown popularly as mixed crop with barley, toria or mustard and intercrop with autumn sugarcane.

4.9 SEEDS AND SOWING

Depending on what subspecies is grown or when sowing is taken up, the seed rate and spacing varies in lentil.

Late sown (mid-November to December end) crop is sown with closer spacing (20–25 cm rows) using 50–60 kg seeds/ha. For normal sowing season (mid-October to mid-November) a seed rate of 30–40 kg/ha is sufficient due to wider spacing (30 cm rows). Bold seeded lentil require relatively higher seed rate by 15–20% than small seeded varieties. In case of utera system, still higher seed rate (80–90 kg/ha) is needed.

4.10 NUTRIENT MANAGEMENT

Although, crop does not generally receive fertilisers, the crop has shown excellent response to fertiliser application. Starter dose of 20–25 kg N/ha + 40–50 kg P_2O_5/ha at sowing time is essential for good crop.

Application of potassium @20–25 kg/ha should depend on the K status of soils. Application of 3–4 t/ha manure along with *Rhizobium* treatment of seeds are also necessary for good yields. In paddy fallows, the crop should receive zinc sulfate application either to soil (25 kg/ha) or as spray (1% solution). This is due to poor status of Zn in paddy soils.

The fertilisers are either broadcasted (in uetra cultivation) to drilled alongside the line, where sowing is done.

4.11 WATER MANAGEMENT

The crop is never irrigated, although, protective irrigations during failure of rains can ensure good yields. Such irrigations are especially necessary in sandy type of soils (having less water holding capacity) at 45 DAS and 80 DAS.

However, the crop must grow in well drained condition. Hence, water stagnation due to rains will have to be avoided.

4.12 WEED MANAGEMENT

As lentil is a slow growing crop, the weed menace may lead to complete crop failure. Hence, 1–2 intercultivations till 60 days of crop period are essential, or hand weeding may also be done. But in the utera crop, the seeds are broadcasted and intercultivation is not possible. Hand weeding may be necessary, if economically viable. In extreme weed intensity, a pre-emergence spray of Pendimethalin @1–1.5 kg a.i./ha or Basalin @0.75 kg a.i./ha may be necessary.

4.13 HARVESTING AND THRESHING

As the shattering is common, the crop must be harvested when most of the pods are ripe and dry without waiting for all the pods to mature. Entire plant is cut and dried thoroughly before pods are separated and threshed. The seeds are further dried to bring the moisture level to 12%. It is possible to expect a yield of around 20–25 q/ha from a good sole crop of lentil.

4.14 DISEASES

- **Seedling mortality:** It is a typical seedling disease. It can be caused by *Fusarium.* Symptoms include either seedling death by yellowed leaves or collar rot and collapse of the seedling. It can be controlled by the seed treatment with Benomyl @2.5 g/kg seed or by delaying the sowing till mid-November.
- **Wilt:** The yellowing, drying and death of seedlings is associated with browning of roots. It can be controlled by the strict crop rotation or by using resistant variety, like DPL 62, Pant L 406.
- **Rust:** Symptoms of the disease (caused by *Uromyces fabae*) include pink or brown pustules on leaves and stem, which later turn into black. It can appear in devastating proportion. It can be controlled by 0.2% Mancozeb spray, 2–3 times, at 15 days interval.
- **Powdery mildew:** (*Erysiphe polygoni*), seen after 2–3 months of crop, the symptoms of this disease include whitish patches on back side of the leaves, which may spread and cover the entire leaf surface, stem and even pods. It can be controlled by spraying 0.3% wettable sulphur two times at 15 days interval.
- **Downy mildew:** (*Peronospora lentis*), causing damages in late stages, symptoms of this disease include light greenish to yellow spots on the upper surface, and lower surface developing brownish cottony growth later becoming chlorotic. It can be controlled by the seed treatment with Carbendazim @2.5 g/kg seed.

No major insect pest has been recorded, although, some polyphagous insects like hairy caterpillar and semi looper may damage the crop occasionally. They can be controlled by the spray of Monocrotophos @2 ml/l.

4.15 ANTI-NUTRITIONAL FACTORS

Although, lentils are considered to be nutritious pulse, it contains many anti-nutritional factors—which reduce the food quality of lentil significantly. Lentils contain Trypsin inhibitors, which effectively reduce the activity of digestive enzyme—Trypsin, and thereby cause digestion problems and reduce the availability of amino acids. They also contain condensed tannins which react with lysine and methionine and reduce their availability during digestion process. In addition, they also contain phytic acids, which very effectively reduce the availability of minerals, by chelating them. These anti-nutritional factors can be reduced by the processes. Cooking soaked seeds remove the trypsin inhibitors, reduce the phytates; Tannins could be removed by germination.

4.16 PRODUCTION TECHNOLOGIES

- **Land preparation:** Well ploughed land is harrowed 2–3 times and levelled to bring the land to good tilth. Very fine seedbed may not be necessary. But, before second harrowing, 4–5 tons of FYM or any other organic manure should be applied.
- **Seeds and sowing:** Seeds are sown by seed drill or plough sole method @ row spacing of 25–30 cm using seed rate 40–45 kg/ha for sole crop. The seed rate may reduce proportionately in intercrops. But in utera cultivation in standing paddy crop, 70–80 kg/ha seeds are broadcasted.
- **Fertilisers:** Fertilisers supplying 25 kg N, 40–50 kg P_2O_5 and 25 kg K_2O per hectare are applied at the time of sowing along with Rhizobium treatment of seeds.
- **Weed control:** As many as 3–4 intercultivations before 60 days are necessary to keep the crop weed free in sole crop sown in rows. In intense weed growth, pre-emergence spray of Pendimethalin @1–1.5 kg a.i./ha may be useful.
- **Water management:** When the crop is grown in rain free environment on light soils, 2–3 protective irrigations may be needed. But, when excess rains occur, drainage must be arranged.
- **Diseases:** Will have to be controlled by the appropriate fungicides.
- **Harvest:** Crop may have to be harvested after 120 days (early varieties) or 130–140 days (mid-early varieties) or 170–180 days (late varieties). A yield of 20–25 q/ha may be expected from sole crop, while 8–10 q/ha may be harvested from inter crop.

REFERENCES

Anonymous (2011), Lentil, www.todayscience.org/AS/v1-1/AS.2291-4471.2013.0101004.pdf.

Anonymous, (2012), Enhancing Lentil Production for Food and Nutritional Security and Improved Livelihoods, NFSM/Presentations/ICARDA23_June_10.ppt.

Anonymous (2014), Lentil, Wikipedia http://en.wikipedia.org/wiki/Lentil.

Anonymous, (2022), Agricultural statistics at a glance, DAFW, GoI, p. 29.

FAOSTAT, (2023), https://www.fao.org/faostat/en/#data/QCL.

Muehlbauer, F.J. and Abebe Tullu (1997), *Lens culinaris* www.hort.purdue.edu/newcrop/cropfactsheets/**lentil**.html.

Valverde, Concepcion Vidal, Juana Frias, Isabel Estrella, Maria J. Gorospe, Raquel Ruiz and Jim Bacon (1994), Effect of Processing on Some Antinutritional Factors of Lentils, *J. Agric. Food Chem*, Vol. 42, 10th ed., pp. 2291–2295.

Chapter 5

Pea (*Pisum sativum* L.)

5.1 ECONOMIC IMPORTANCE

Immature green seeds of peas are a popular vegetable throughout the world. Often, they are frozen or canned before use. All such peas are called **green peas**. Mature dry grains, either whole or split (called **dry peas**), are also consumed as rich source of protein, in different recipes. Pea is also used as green manure or as rich cattle feed (both as green and dry) or even used as cover crop to reduce soil erosion.

It is popularly called **matar** in Hindi and many other names in vernacular languages. Like many other pulse crops, it is known to improve the soil fertility due to N fixation. Due to its short duration, it can fit into many sequential cropping systems. Its dry grains are known for low fat (1%), high protein (25%) and medium in carbohydrates (60%), besides vitamins like Thiamine (B_1), Pantothenic acid (B_5) and folate (B_9). It is also rich in minerals like calcium and iron. But raw green peas, used as vegetable, contain lesser carbohydrate (15%), lesser protein (5%) than dry grains.

5.2 ORIGIN AND HISTORY

Peas was originated in the **Mediterranean region of the South Europe**. It later spread to Western Asia, Afghanistan (2000 BC), North Western India (1700 BC) and China. By 200 BC, it had been occupied as an important crop of Gangetic belt in India. It has also entered temperate countries like Russia as early as 500 BC.

5.3 CLASSIFICATION

Peas are classified in two groups:

1. **Garden pea (*Pisum sativum* var. *hortense*):** It is also called **table pea**. Young green immature seeds are harvested for vegetable purposes either used in unprocessed or in canned/frozen condition. Seeds are bold and wrinkled, when dried. They look fleshy round green when immature. Dry seeds are green or yellowish or bluish green. The plants have white flowers.

2. **Field pea (*Pisum sativum* var. *arvense*):** It is also called **dry peas**. Ripe mature seeds are used as whole or split seeds. Dry seeds are small rounded (unwrinkled seed coat) grayish green, grayish brown or grayish yellow. Plants have varying coloured flowers. They are more hardy plants, with an ability to withstand frost and drought than garden peas.

Why Vegetable Peas are not used as Field Peas?
Garden peas, when dried, has dry wrinkled seed coat and has lower protein content than field peas. Hence, a garden pea is always used as vegetable–in fleshy, roundish plumpy condition, although, its protein is lesser than dry field peas. Similarly, young immature seeds of field pea are not preferred as vegetable, as they are not **fleshy** and **plumpy**.

5.4 AREA AND DISTRIBUTION

It is produced on 7.4 million hectare world over, with a production of 13.76 million tons (average global productivity 1857 kg/ha). Large area under peas is found in China, Russia, USA and India. In India, it is grown over 8.4 lakh hectare with a production of 11.3 lakh tons. Around 25% of the pea production is consumed as green peas. About two-third of the Indian production comes from Uttar Pradesh, while remaining states like Madhya Pradesh and Bihar and parts of Punjab also produce substantially. All remaining states, including South India, have negligible area.

5.5 SOIL AND CLIMATIC REQUIREMENTS

Well drained loamy soils with pH range of 6.0 –7.5 are most congenial to grow pea. **Sandy loams**, and **silty loams** may be suitable, but clayey soils are not suitable if the drainage is a constraint. It is highly sensitive to even small content of salts in soil. Hence, it cannot be grown in any type of saline soils.

Field pea is successful in **temperate** and **subtropical** regions. It is more successful in temperate semi-arid conditions. For germination, ideal temperature is 22°C, while for further growth, ideal temperature vary between 13–18°C. But late season frost can damage the crop (especially flowering), while it may withstand early frosts. But, damage by drought can be more severe than frost. Higher humidity can always accentuate diseases. Temperature > 30°C can reduce pollination. It grows well above 750 m M.S.L. Ideally, it requires 500–550 mm rainfall during its crop period (120–140 days).

The growing degree days required for pea till maturity is in the range of 1520–1670, indicating the cold loving nature of the crop, out of which GDD required till grain development is 960.

5.6 BOTANICAL DESCRIPTION

It is an annual herbaceous semi-erect plant, with a tendency to climb on any support, growing to a maximum height of 2 m. It belongs to *Fabaceae* family.

Stem is tender, hollow, and succulent, while leaves are pinnately compound (1–3 pairs of leaflets/leaf) ending with branched tendrils (Figure 5.1). Base of petiole has large leaf like bract, often confused for sessile leaf. Inflorescence is axillary raceme with a typical papilionaceous self-pollinated flowers. Fruit is 5–9 cm pod with 4–9 seeds (Figure 5.2), which may be round or angular, greenish yellow or grey.

Figure 5.1 Tendrils of pea plant.

Figure 5.2 Pods of garden peas.

5.7 VARIETIES

Several varieties of field pea and garden pea are released for diversified climatic situations of the country. Most of them show local adaptability, although, temperature decides the success of crop. Some important varieties of pea recommended in different states are listed in Table 5.1.

Table 5.1 List of Pea Varieties Recommended in Different States

State	*Varieties Recommended*
Assam	HUP-2, Retina, Aparna
Bihar	KFP 103, HUP-2, Uttara, Sikha, Aparna, Retina
Haryana	Alankar, DMR 11, Uttara, Jayanthi, Aparna
Himachal Pradesh	Aparna, DMR 7, DMR 11
Punjab	PG 3, Aparna, DMR 11
Rajasthan	HUP-2, Retina, Aparna
Uttar Pradesh	HUP-2, Retina, Aparna, Sikha, Uttara, Pant Pea 2, JP 885, Malvia, Swathi, Sapna
Uttarakhand	VL Pea 1, JP 885, Retina, DMR 11
Madhya Pradesh	Aparna, JP 885, Retina, DMR 11
West Bengal	Aparna, HUP 2, Sikha, Retina, Uttara, KFP 103

5.8 CROPPING SYSTEMS

It is popularly grown after *kharif* crops like maize, rice, cotton, sorghum or pearl millet in sequential systems under irrigation, to ensure that pea is exposed to cold temperatures. It is also grown as mixed crop with crops like chick pea, barley, wheat, oats, rape seed and mustard crops, either for vegetable purposes or for dry grains.

It is also grown as intercrop with number of *rabi* crops with inter row spacing of 30 cm in between wide spaced crops like sugarcane and cotton. Pea neither dominates on other major crops, nor it is sensitive enough to be affected by them.

5.9 NUTRIENT MANAGEMENT

It is not a heavy feeding crop. Generally, its N requirement is less, due to biological N fixation. However, basal dose of N dose of 20–30 kg/ha is necessary till nodulation starts. Excess N application may promote more tendril formation and may reduce the yield. The crop requires more phosphorus (dose: 60 –70 kg/ha) to promote netter nodulation. Basal dose of about 30–40 kg K_2O per hectare is also recommended. When the soil is deficient in zinc, application of zinc sulphate (10–12 kg/ha) is useful.

5.10 WATER MANAGEMENT

Pea is normally grown as rainfed crop. Despite being grown as un-irrigated crop, it responds very well, if irrigated. Two irrigations, one at 45 DAS and other at pod filling stage are essential to increase yield.

5.11 WEED MANAGEMENT

The crop must be weed free at least up to 45 days to avoid competition by weeds. This is achieved by 1–2 intercultivations. Application of Basalin @0.75 kg a.i./ha or Metribuzin @1–1.5 kg a.i./ha as pre-sowing spray can be used safely, to reduce the weed infestation.

Important Weeds in Pea
Chenopodium album, Parviflora, Lathyrus spp., Melilotus alba, Vicia faba

5.12 PESTS AND DISEASES

- **Wilt and root rot:** In early sown crop, these fungal diseases are expected at seedling stage. The symptoms include yellowing of lower leaves followed by wilting. They can be controlled by the seed treatment with Carbendazim (@2.5 g/kg seed) or avoiding early sowing.
- **Powdery mildew:** White powder patchy growth are seen on both sides of the leaves and later spread to other parts, including tendrils. In severe conditions, the plant may die. It can be controlled by the spray of 3 kg wettable sulphur in 1000*l* (in two sprays of 15 days interval).
- **Rust:** Yellow spots appear on all green parts initially, which later develop into powdery brown appearance. In severe cases, the plants may die. It can be controlled by the spray of Mancozeb @2 g/l.

- **Stem fly:** Occurring in early plantings, the maggots damage internal tissues to lead to the death of plant in early stages of the crop. It can be controlled by the use of 30 kg Carbofuran/ha or 10 kg Phorate/ha applied to soil.
- **Pea aphids:** They suck sap to turn the leaves pale and yellow to arrest the growth. It can be controlled by the spray of Metasystox or Dimethoate (@1.25 ml/l).
- **Pod borer:** Prevalent in late sown crop, the infestation is indicated by a hole on the pod and caterpillar feeding on young seeds. It can be controlled by the use of Cypermethrin (1.25 ml/l).

5.13 ANTI-NUTRITIONAL FACTORS

Peas contain significant levels of anti-nutritional factors such as phytic acid, total phenolic acids, and trypsin inhibitors. Phytic acid has chelating effect on minerals and reduce their bioavailability. Phenolic acids can decrease the protein accessibility to humans. Trypsin inhibitors can decrease the protein utility by inactivating the digestive enzyme, trypsin. These anti-nutritional factors could be significantly reduced by the processes like heat treatment (cooking), germination or simple soaking.

5.14 PRODUCTION TECHNIQUES

- **Land preparation:** One or two ploughings followed by one harrowing is sufficient to prepare the land. Application of 10 tons FYM/ha is desirable before final harrowing at least 15 days in advance of sowing. Then the field should be levelled, before sowing.
- **Seeds and sowing:** Treatment of 60–80 kg/ha seeds with Rhizobium (@600–700 g/ha) as well as Thiram or Carbendazim (150 g/ha) is necessary. Seeds are sown in the rows of 30 cm by seed drill or plough sole method in the first fortnight of November. Early varieties are given closer inter row spacing (20 cm) and use higher seed rate (up to 110 kg/ha), while late maturing varieties are sown at wider spacing (30 cm) using relatively sell seed rate of 60 –70 kg/ha. Time of sowing should be synchronised to time when temperatures are in the range of 20–30°C to avoid wilt disease.
- **Fertilisers:** Fertilisers supplying 20–30 kg N, 60–70 kg P_2O_5 and 30–40 kg K_2O per hectare are necessary to be supplied as basal dose. They are placed in rows alongside of crop rows.
- **Irrigation:** The crop is hardly irrigated. However, if irrigated, two irrigations (first at 45 DAS and second at pod filling stage) are given. Water logging has to be avoided.
- **Plant protection** Plant protection measures are expected to be followed, on need basis and prophylactic measures are necessary when sowing time is preponed or postponed.
- **Harvest and yield:** Harvesting is done by picking green and immature pods for vegetable purposes or by cutting the entire plant for dry pea purposes. They are thoroughly dried and pods are separated and threshed by beating. 20–25 q of dry peas are expected from one hectare.

REFERENCES

Anonymous, *Garden Pea Production*, Department of Agriculture, Forestry and Fisheries, Republic of South Africa, Vol. 99, pp. 3–30, 2007.

Anonymous, 2022, Agricultural statistics at a glance, DAFW, GoI, p. 29.

Chen, Y.M., Wang, K., Maskus, H., Bourré, L., Malcolmson, L. and Arntfield, S., *Effects of Processing on Anti-Nutritional Factors in Yellow Pea Flours and Their Pizza Dough Products*, Paper presented at University of Manitoba, Canada www.uwa.edu.au/_data/assets/ pdf_file/0008/85526/Pulse-improvement-in-WA.pdf+&cd=5&hl=en&ct=clnk&gl=in, 2011.

FAOSTAT, 2023, https://www.fao.org/faostat/en/#data/QCL.

Miller, Perry, Will Lanier and Stu Brandt, *Using Growing Degree Days to Predict Plant Stages*, Montguide, 1995.

Muehlbauer, F.J. and Abebe Tullu, *Pisum sativum,* www.hort.purdue.edu/newcrop/cropfactsheets/**pea**.html, 1997.

Chapter 6

Rabi Red Gram (Pigeon pea) (*Cajanus cajan* L.)

6.1 ECONOMIC IMPORTANCE

Pigeon pea is also known as **red gram** or **tur** or **arhar** or **glandule pea** or **congo pea**. It is the most important staple source of protein in daily food, popularly used as 'dal' or 'tur dal' along with other cereal products. It contributes 15% of total pulses from 20% of the pulse area in the country. It is consumed as **dry grain** after dehulling and splitting into two halves; or as green **unripe grain** or even young leaves to make variety of culinary preparations. It is a rich source of protein (21.7%)—which is made up of tryptophan, tyrosine, cysteine, arginine, methionine, lysine, carbohydrate (63%), calcium (130 mg/100 g) magnesium (183 mg/100 g), phosphorus (367 mg/100 g), potassium (1.4 g/100 g), besides all B types of vitamins. But, red gram is devoid of other vitamins and is low in fats (1.5%). It is also a rich cattle feed (husk of the pod, husk of the grain, dry/green leaves).

It is a highly drought resistant crop and hence a preferred crop in many semi-arid tracts. Depending upon its duration, it fits into many cropping systems. It is well known to fix atmospheric nitrogen symbiotically in association with *Rhizobium*; it can also be used as green manure (up to 90 kg/ha N may be available). Its hard stem is a popular source of fuel in villages. It has also been used as a host to rear lac yielding insects.

6.2 ORIGIN AND HISTORY

It originated in the Eastern part of the **peninsular India**, as evidenced by large diversity in its forms, the presence of wild forms (closest wild form, called ***Mansi*** is found in tropical deciduous forests in Odisha) and archaeological evidences (Neolithic sites of Odisha). It must have travelled to other parts of the world (mainly Southern Asia, Central Africa and Central America) from India many centuries ago.

6.3 AREA AND DISTRIBUTION

Pigeon pea is grown over 5.35 million hectare in three continents viz., Asia, Africa and Central America—in more than 25 countries of the tropical and subtropical climates, with a global productivity of 851.1 kg/ha. Pigeon pea is principally grown in India (40.7 lakh hectare, contributing 76% of the global area), Myanmar (4.1 lakh hectare), Mozambique, Malawi (2.56 lakh hectare), Tanzania (67000 hectare), Kenya (3.00 lakh hectare) and Uganda (40000 hectare) and on a limited scale in some Caribbean countries of Central America, besides Australia and Sri Lanka. (FAOSTAT, 2009)

In India, it is mainly grown in the states of Maharashtra (1.34 million hectare; 1.37 million tons @ 1023 kg/ha), Karnataka (1.72 million hectare; 1.14 million tons @ 666 kg/ha), Madhya Pradesh (0.37 million hectare; 0.35 million tons @ 936 kg/ha), Uttar Pradesh (0.29 million hectare; 0.35 million tons @ 1196 kg/ha) and Gujarat (0.25 million hectare; 0.29 million tons @ 1160 kg/ha). Other states contribute negligible area and production.

6.4 CLASSIFICATION

- ***Cajanus cajan* var. *bicolor*:** Late maturing, tall growing upright bushy plants, with **yellow flowers** and **purple streaked pods**, each with 4–5 seeds. Its **standard** (largest petal) has **red veins**.
- ***Cajanus cajan* var. *flavus*:** Early maturing, smaller spreading bushy plant, with yellow flowers having **no streak on standard** and having **plain green pods**, each having 2–3 seeds.

6.5 SOIL AND CLIMATIC REQUIREMENTS

Well-drained, deep, clay loam soils are most suitable to grow red gram. However, it also grows on alluvial soils, sandy loams as well as deep black cotton soils. Its performance predominantly governed by **depth** and **drainage of soil**, besides fertility, irrespective of texture. It cannot withstand salinity and water logging, as they may adversely affect root growth. Ideal pH 6–7.5.

It is a **short day plant**, and flowers well in shorter days. Hence, sowing date should be adjusted to coincide with the short days of at least < 12 hours of a given locality. Both photoperiod and genotype decide the duration of crop (100 days to 240 days). Longer photoperiod during flowering stage can prolong the duration.

It requires **warm humid climates till flowering** and **dry hot climates during post flowering** phase. In India, this requirement for vegetative growth is met between June–October and for reproductive growth it is met between November and January in most areas. Ideal requirement of temperatures are: germination: 20–25°C; branching: 22–30°C; flowering: 16–21°C; and post flowering: 28–38°C. Lower temperature may limit the germination and early growth. But, maturity/grain development is not affected by high temperature, if moisture is not a limitation. Pod development stage requires clear sunny days, as cloudy skies may retard their growth. It is highly susceptible to frost and temperature below 12°C.

Rainfall requirement of red gram varies according to the duration (500–950 mm). But, it can sustain the drought during post flowering period, while pre-flowering period requires more rainfall or irrigation (water requirement = 40–60 cm).

6.6 BOTANICAL DESCRIPTION OF PLANT

Red gram is an **annual** or **perennial shrub** belonging to *Fabaceae* with chromosome number $2n = 22/44$ (diploid and tetraploid) with profuse branching habit. Plant height varies from 1 to 4 m.

- **Roots system: Strong tap root**, with primary and secondary (laterally grown) branches. Spreading types have shallower root system, while upright types have deeper root system.
- **Leaves:** Compound, trifoliate with entire leaflets, silky and densely hairy on lower sides.
- **Inflorescence:** It is an **axillary raceme**, bearing papilionaceous flowers, with two flowers opening at time on one raceme (Figure 6.1). Flowering continues till harvest; usually self-pollinated (pollination before opening of flower), but cross-pollination may occur up to 20%. In late varieties, flowers are grouped together at the ends of branches, but in early variety, flowers appear in all axils of branches.
- **Pods:** Green or dark brown, 5–10 cm long, streaked/plain bodied having pointed tips (Figure 6.2).
- **Seeds:** Size, shape and colour of seeds vary greatly. Usually, they are round and lens shaped.

Figure 6.1 Trifoliate leaves and flowers of pigeon pea.

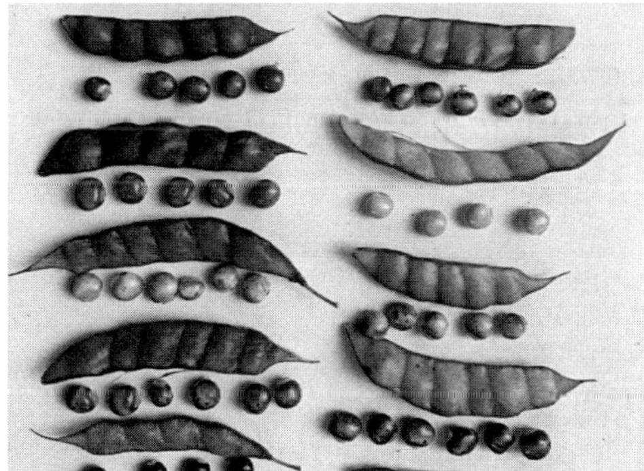

Figure 6.2 Dried pigeon pea pod.

6.7 VARIETIES

Many red gram varieties and hybrids are developed to suit the diversified agro-climatic situations of red gram growing areas of the country. Most of them are location-specific and photo-sensitive. Hence, their duration may vary vastly across the country. Some prominent varieties are listed in Table 6.1.

6.8 CROPPING SYSTEMS

Pigeon pea has diversified spacing and duration, besides its N fixing abilities for making it an ideal crop for a large number of sequential and intercropping systems. However, long duration pigeon pea is not followed by the second crop, especially in the rainfed situations.

Table 6.1 List of Important Varieties and Hybrids of Red Gram

States	*Recommended Varieties and Hybrids of Red Gram*
Andhra Pradesh	LRG 38, LRG 41, LRG 30, GTH 1, ICPL 88039, PRG 158, Sarita, Asha, Durga, Laxmi, PRG 100, ICPH 2671 (hybrid)
Bihar	MAL 13, MAL 6, Azad, Birsa Arhar 1
Delhi	Pusa 991, Pusa 885, Asha, Birsa Arhar 1
Gujarat	GTH 1, ICPL 88039, GT 101, ICPL 151, GT 101, AKPH 4101 (hybrid), Jawahar, Asha
Haryana	ICPL 88039, GTH 1, UPAS 120, Pusa 992/855, ICPH 8 (hybrid), ICPL 151
Karnataka	Sel-31, TTB7, Hyd-3C, BRG 1/2, ICPL 7035
Maharashtra	ICPL 88039, GTH 1, UPAS 120, AI 15, AL 201, GT 101, ICPL 151, Vaishali, AKT 8811, MA 3, AKPH 4101 (hybrid), BSMR 736, Jawahar, Sharad, ICPH 8 (hybrid)
Madhya Pradesh	ICPL 88039, GTH 1, ICPL 88039,GT 101, MAL 13, Manak, MA3, ICPL 151, AKPH 4101 (hybrid), Jawahar, JA 4, ICPH 8 (hybrid),
Odisha	Asha, ICPH 8 (hybrid),
Punjab	GTH 1, ICPL 88039, ICPL 88039, PPH 4 (hybrid), al 201, Pusa 855, Birsa Arhar 1
Rajasthan	GTH 1, ICPL 88039, UPAS 120, Pusa 992, Pusa 855, Birsa Arhar 1
Tamil Nadu	Co RG 9701, Co PH 2, Co-6, UPAS 120, Asha, Vamban 1
Uttar Pradesh	ICPL 88039, GTH 1, UPAS 120, Pusa 992, Pusa 855, Birsa Arhar 1 MA 6, Manak, Azad, Amir, NDA 82-2, Pusa 9
West Bengal	Pusa 992, Pusa 9

Depending upon the duration of the crop, it is followed by *rabi* crops like wheat, potato, lentil, sugar cane or wheat followed by green gram (in 3 crop sequence under irrigated).

By adjusting spacing, it is popularly grown with crops like green gram, cowpea, and black gram in intercropping system. Being deep rooted crop, it is also grown as intercrop with shallow rooted ragi.

But, exhaustive crops like cotton, maize or crops aggravating red gram pests (like castor) should be avoided as intercrops.

6.9 SEED RATE AND POPULATION DENSITY

Seed rate and population density are higher, as the **duration of the crop** reduces because the growth of the short duration crop is less vigorous than late maturing crop. Similarly, **early sown** crop (of any variety) requires less population (or less density) than **late sown** crop.

Long duration crop (> 180 days):

- 120 cm × 30 cm or 90 cm × 45 cm (seed rate of 10–12 kg/ha for a density of 24700–27777/ha) for heavy soils.
- 90 cm × 30 cm (seed rate of 15 kg/ha for a density of 37000/ha) for light soils.

Medium duration crop (150–180 days):

- 75 cm × 30 cm (seed rate of 15–16 kg/ha for a density of 44444/ha) for heavy soils.
- 60 cm × 30 cm (seed rate of 16–18 kg/ha for a density of 55555/ha) for light soils.

Early duration crop (100–150 days):

- 45 cm × 20 cm (seed rate of 20–22 kg/ha for a density of 111111/ha).

6.10 SEASONS OF CULTIVATION

Traditional as well as improved long duration varieties should be grown on the **longest day** (during the month of June–July) to avoid reduction in vegetative growth and resultant poorer yield caused due to late sowings. However, it is now recommended in North India to sow the crop by April (if irrigation is available) so that the second crop of the wheat could be accommodated. But, it is possible in such areas, where short days would commence early during September.

Early maturing varieties (100–140 days) could be sown up to the end of July without yield reduction, where the winters are milder (Peninsular India and Eastern Uttar Pradesh). A short duration *rabi* red gram can be grown (sown during September–October), wherever the short days could be experienced till January. But, its yields may be reduced, if the crop experience stress during post flowering phase, especially under rainfed conditions.

6.11 NUTRIENT MANAGEMENT

Application manures @5–7 t/ha at the time of land preparation is essential for the successful crop, although the farmers ignore applying the manures. Whenever red gram is grown as an intercrop, no fertiliser is applied to red gram (assuming the dose applied to main crop is available to red gram), resulting into reduced productivity. Pigeon pea is mainly dependent upon **symbiotic nitrogen fixation** through induced nodulation. However, the response to artificial inoculation depends upon the use/availability of the proper strain of *Rhizobium*, soil moisture, soil organic carbon and method of application.

The crop does need a **starter dose of nitrogen** (20–25 kg/ha) for initial needs, which may be unnecessary in high fertility situations. Later N needs of the crop are met by biological nitrogen fixation. Higher **initial dose of N** may lead to **disproportionate vegetative growth** (leading to low yield) or may not result in incremental yield.

Response to the applied phosphorus is up to 60–80 kg/ha, especially in P deficient soils. Phosphorus is required to **promote the nodulation** and **nitrogen fixation**. Even in medium P soils, a dose of 20–30 kg/ha of phosphorus is generally necessary. The response to addition of K fertilisers is inconsistent. K fertilisers are applied to supply at least 20 kg/ha, especially in K deficient soils.

Although, the crop needs sulphur, its demand can be met by the use of super phosphate as P source. In Zn deficient soils, additional application of $ZnSO_4$ (20 kg/ha) may be necessary.

6.12 WATER MANAGEMENT

Being a drought resistant deep rooted crop, it is hardly grown under irrigation. However, sole crops of red gram are protectively irrigated (2–3 during reproductive stages). Due to its sensitivity to poor drainage, care should be taken to make sure that ridges and furrow method is followed.

Even under rainfed conditions, **incessant rains may not be conducive** to the crop. Prolonged drought beyond 30 days may affect the crop. Its water requirement varies from 40 to 50 cm, depending on the duration.

6.13 WEED MANAGEMENT

Initial **60 days are critical** to weeds. But, control of weeds is easier, as intercultivation is possible in wide spaced crop like red gram (20 and 45 days after sowing), which may be supplemented with hand weeding, if necessary. In case of severe weed infestation, Alachlor (@2.5 l/ha) or Basalin (@1 kg a.i./ha) as pre-emergence spray is able to control most of the weeds till 60–70 days.

6.14 HARVEST AND POST-HARVEST

Harvesting can be made by cutting the entire plant at the base after at least 75% pods are **dried/turned brown**. The cut plants are left on the field for drying for 1–2 days. Then the pods are separated and beaten to separate the seeds or the whole plant is subjected to beating. Mechanised threshers are also used in regular pigeon pea areas. Pod: Seed ratio is 1:0.6 and seeds to stubble ratio are 1:2.5. Seeds are dried up to 11% before they are subjected to dal making process.

A grain yield of 20–25 q/ha along with stubble of 50–60 q/ha may be expected from a good sole crop, while as an intercrop, its yields may be in the range of 6–10 q/ha.

6.15 DISEASES AND PESTS

- **Wilt:** Leaf yellowing followed by the drying of whole plant as well as the streaks beneath the epidermis of the main roots are the symptoms of wilt caused by *Fusarium oxysporium*. Affected plants can be easily pulled. It can be controlled by avoiding mono-cropping as well as intercropping with jowar or by using disease resistant varieties, like Maruthi, Azad, Amar or Asha.
- **Stem rot:** Dark brown lesions on stem near soil surface, ultimately leading girdling and drying of whole plant are the symptoms of this disease, caused by *Phytophthora dreschsleri*. Roots are intact and hence cannot be pulled. It can be controlled by providing good drainage or by the use of resistant varieties.

- **Sterility mosaic:** The symptoms of disease are spread by the vector *Eriophyd* mite and include profuse branching, reduced leaf size and bushy appearance with no flowering/ fruiting. The vectors can be controlled by the spray of 1% Metasystox for 3–4 times.
- **Pod borer:** Deadly and widely spread pod borer (*Heliothis armigera*) can damage the whole crop. Symptoms include prominent hole on the pod with a green caterpillar typically feeding by entering the half body. Even tender leaves/twigs may be eaten. It can be controlled by the spray of Monocrotophos (@1.5 ml/l) alternated with pyrethroids @0.5 ml/l.
- **Podfly:** Maggots feed on seeds and pods may get deformed. It can be controlled by the spray of Monocrotophos @1.5 ml/l.
- **Plume moth:** The caterpillar feeds on developing seeds or flowers or even buds, making them to drop. It can be controlled by the spray of 1% Metasystox.
- **Leaf hopper:** Both adults and nymphs suck the sap from the lower surface of the leaves resulting in loss of turgidity, turning the leaf into brown colour with typical edge curling. It can be controlled by the spray of Monocrotophos @1 ml/l.
- **Galerucid beetle:** It attacks after dusk, concealing itself under the soil during day hours. The damage includes typical circular holes on the leaves. It can be controlled by application of Phorate to soil @10 kg/ha.

6.16 PRODUCTION TECHNOLOGIES

- **Land preparation:** A deep ploughing followed by 2–3 harrowing, and then land is laid out in ridges and furrow after manure application (@5–7 t/ha). Fall ploughing is more useful.
- **Season:** First week of July (normal); first to third week of June (early); September-October (*rabi*).
- **Sowing:** By dibbling at inter row spacing of 90 cm (light soils) to 120 cm × 30 cm (heavy soils) or 60–90 × 30 cm for medium duration and 45 × 20 cm for very early varieties; seed rate 10–22 kg/ha, depending on spacing. Seed treatment with *Rhizobium* before sowing is essential.
- **Fertilisers:** 20–25 kg/ha N; 40–50 kg P_2O_5/ha; 20 kg K_2O/ha. All fertilisers are applied as basal dose; application of $ZnSO_4$ @20 kg/ha along with basal dose is useful.
- **Irrigation:** It may be given in sensitive stages like flowering and pod development stages.
- **Weed management:** Intercultivation at 20 and 45 DAS along with pre-emergence spray of Alachlor @2.5 l/ha.
- **Plant protection:** As suggested, depending on the incidence of pests and diseases.
- **Harvesting and yield:** By cutting entire plant at base, when 75% of pods are dry and leaving them in field for 1–2 days. Later the pods are separated by beating and then are subjected to mechanical threshing to separate the seeds. They are dried under the sun till the seed moisture reaches 11%. A yield of 20–25 q/ha along with 50–60 q/ha stubble may be expected from the main crop and as intercrop, it may yield as low as 6–10 q/ha.

6.17 ANTI-NUTRITIONAL FACTORS

Like many pulses, red gram has anti-nutritional factors such as **phytolecithins**, **polyphenols** and **oligosaccharides**. Polyphenols (tannins) inhibit the digestive enzymes like protease, amylase, trypsin and chymotrypsin and reduce the digestion of protein and carbohydrates when red gram is consumed wholly, without removal of hull. Polyphenols may also make the minerals unavailable by binding them. Phytolecithins contained in pigeon pea dal may agglutenise glycol protein on the surface of red blood cells. But, lecithins are heat sensitive and are removed during cooking. Half of the sugar contained in the pigeon pea seeds are in the form of oligosaccharides, like stachyose, raffinose, and verbascose. They may cause flatulence, when consumed in large quantity.

REFERENCES

Anonymous (2009), FAOSTAT, http://faostat.fao.org/site/424/default.aspx#ancor.

Anonymous, Indian Production Centres, Empowering Agriculture, Food Processing and Food Safety http://www.efreshglobal.com/eFresh/Content/Products.aspx?u=Redgram_pcerind.

Anonymous, Pigeon pea, ICRISAT, http://www.icrisat.org/crop-pigeonpea.html, http://efreshindia.com/eFresh/Content/Products.aspx?u=Redgram_pcerind.

Anonymous, 2022, Agricultural statistics at a glance, DAFW, GoI, p. 29.

FAOSTAT, 2023, https://www.fao.org/faostat/en/#data/QCL.

Singh, Faujdar and B. Diwakar (1993), *Nutritive Value and Uses of Pigeonpea and Groundnut*, ICRISAT, Hyderabad, pp. 8–9.

Yadav, D.S. (2005), *Pulse Crops*, Kalyani, Ludhiana.

Chapter 7

Rajmash (*Phaseolus vulgaris* L.)

7.1 ECONOMIC IMPORTANCE

It is a popular vegetable crop in India and the most widely cultivated edible pulse vegetable throughout the world, especially in tropical and subtropical regions. It is called by different names, like **French beans, common beans**, **kidney beans**, **snap beans**, **string beans** and even **wax bean**. Besides using immature pods as vegetable, its dry mature grains are also consumed as pulse for different culinary preparations. Dry grains are called **Rajma** in India. It is a popular pulse consumed in many parts of India, especially in Uttar Pradesh, Punjab and Jammu & Kashmir. The vegetable immature pod is rich in fibres, besides containing 1.7% protein, vitamins A and C as well as calcium. But, dry beans (also called **rajma beans**) are rich in protein (up to 23%).

7.2 ORIGIN AND HISTORY

Genus *Phaseolus* has large number of species, out of which largest cultivated species is *vulgaris*. Other cultivated species of *Phaseolus* are *P. coccineus* (runner scarlet bean), *P. lunatus* (Lima bean) and *P. acutifolius* (tapery bean). French bean originated in **Central America**. In India, French bean is most popular, both as a vegetable and pulse grain.

7.3 SOIL AND CLIMATIC REQUIREMENTS

Rajmash is a day neutral **cool season** vegetable crop, which can tolerate high temperature better than peas. Optimum monthly temperature for the cultivation of Rajmash is 15–25°C compared to 10–18°C for peas. It is **sensitive to high rainfall**, **frost** and **very high temperatures**. Pole types tolerate high rainfall better than bushy varieties. Soil requirements are same as that of pea. Ideal soil pH for the growth of French bean is 5.5–6.0.

7.4 BOTANICAL DESCRIPTION OF PLANT

Rajmash is of two types: **bushy** and **climbing**. Bushy plants are grown without support at closed spacing, while climbers are grown with the support of pole. Hence, they are called

pole types. French bean has tap root system with **poor nodule formation**. Leaves are trifoliate. The unripe pods are green in colour (Figure 7.1), while the mature dry seeds are of red/waxy colour (Figure 7.2). Though a self-pollinated crop, French bean offers wide variability in respect of plant growth (bushy/climbing), colour of pod (green/waxy coloured), cross section of pod (flat/oval/round), pliability (stringed/string less) etc.

Figure 7.1 Unripe long pods of french bean.

Figure 7.2 Mature dry seeds of French bean.

7.5 VARIETIES

Specific snap bean varieties are developed for specific purposes. The processing (canning) types of varieties are very popular in the USA, while, varieties used for fresh cooking are popular in India. Some important varieties recommended and grown in India are:

- **Arka Komal:** Introduced bushy variety from Afghanistan. Pods are straight, flat and green with large brown seeds. Good transport and keeping quality. Yield 19 t/ha or 3 t/ha dried seed in 65–70 days.
- **Arka Subidha:** Plants are bushy and photo-sensitive. Pods are straight and oval, light green, string less and fleshy. Yield 19 t/ha in 70 days.
- **Contender:** Released from IARI Regional station, Katrain. Plants bushy with pink flowers. Pods are green, round, long and string less. Tolerant to mosaic and powdery mildew. Yield 20 t/ha.
- **Pusa Parvathi:** Developed through irradiation followed by the selection from wax podded variety EC 1906. Plants bushy with pink flowers. Pods are green, round and long. Resistant to mosaic and powdery mildew. Yield 22–25 t/ha.
- **Pusa Himalatha:** Pole variety with medium sized (14 cm long) round, meaty, string less pods with an average yield of 26 t/ha.
- **VL Boni 1:** Released from VPKAS, Almora, Dwarf variety with white flowers. Pods are round, light green, string less and fleshy. First harvest 45–60 DAS. Yield 10–11 t/ha.

- **Ooty 1:** Released from TNAU, Coimbatore, moderately resistant to leaf spot, anthracnose and pod borer. Yield 10–11 t/ha.
- **TKD 1:** A pole type suitable for growing in hills. Pods are long, and flat with low fibre. Yield 5–6 t/ha in 90–100 days.

Note: The yield of vegetable pods is expressed.

7.6 PRODUCTION TECHNIQUES

- **Land preparation:** Land is ploughed to a fine tilth and divided into plots of convenient size. Ridges and furrows are made by ploughing after application of farmyard manure @15–20 t/ha. Field is irrigated and seeds are sown under optimum moisture condition on one side of the ridges, 2–3 days after irrigation.
- **Plant population:** Early varieties are sown at a spacing of 45–60 cm between rows × 10–15 cm between plants; seed rate required is 80–90 kg/ha. Pole types long duration varieties are sown at 1.0 m apart in hills @3–4 plants/hill and seed rate is much less (25–30 kg/ha).
- **Season:** In the plains of North India, French bean is sown during two seasons viz., July– September and January–February. In peninsular India it is sown in all three seasons. In the hills, sowing is done from March to May.
- **Manure and fertilisers:** A fertiliser dose supplying 50 kg N, 75 kg, P_2O_5 and 75 kg K_2O per hectare are recommended. For pole types, half of N along with full P and K should be applied as basal dose at the time of making ridges and furrows or one or two weeks after germination. Remaining dose of N is applied one month after first application. For bush types, entire fertiliser dose of fertilisers is applied as basal dose.
- **After care:** French bean is a shallow rooted crop and only light inter-cultural operations are practiced. During early stages of crop, weeding followed by fertiliser application and earthing up can be synchronised. A pre-sowing application of Fluchloralin @2 l/ha checks weed growth for 20–25 days.

7.7 WATER MANAGEMENT

Water stress influences yield of French bean and crop is most sensitive at flowering and fruiting stages. 3–4 irrigations are required during growing season for pole types, while bush types are irrigated 1–2 times.

7.8 STAKING

Staking is an important operation for pole types and bamboo sticks or any locally available materials should be erected when plants start vining. Individual vertical stakes and horizontal canes at 40 cm distance are erected for encouraging growth and spread of plants.

Application of plant growth regulators like PCPA (2 ppm) and NAA (5–25 ppm) has favourable effect on fruit set and yield.

7.9 PESTS AND DISEASES

Crop is affected by pests like stem fly, thrips, mites, bean beetle, bean weevil, aphids, etc. Yellow mosaic, anthracnose, powdery mildew, rust, root rot and wilt and leaf spot are common diseases affecting French bean.

7.10 HARVESTING AND YIELD

Pods are harvested at full grown stage. But immature and tender pods are ready for harvest 7–12 days after flowering, depending on the varieties. In bush varieties, 2–3 harvests and in pole types 3–5 harvests are made. Quality of beans varies with harvests.

REFERENCES

Ahlawat, I.P.S. and Shivakumar, B.G. (2005), Agronomy of Kharif and Rabi Crops, ICAR, New Delhi.

Anonymous (2019), Handbook of Agriculture (6th Edition), ICAR, New Delhi.

Chaudhary, B.D., and Sharma, R. (2014). Legume Crops, Agrotech Publishing Academy, Udaipur.

Singh, S.S. (2017), Crop Production: Principles and Practices, Kalyani Publishers, New Delhi.

Yadav, D.S. (2011), Pulse Crops, Kalyani Publishers, Ludhiana.

MODEL QUESTIONS AND KEY ANSWERS FOR RABI PULSES

Choose Most Appropriate Answer and Fill in the Blanks

1. Small lens-shaped seeds called microsperma of lentil are grown in .... (Europe/Mediterranean area/India/Turkey)
2. Cotyledons of lentil are................ (deep yellow/yellowish red/reddish orange/purplish red)
3. Scientific name Lens culinaris refers to .... (field bean/lentil/peas/chick pea)
4. .... country does not grow peas. (Russia/USA/Mexico/India)
5. Growing degree days of peas is as small as .... (1100–1150/550–650/1500–1600/1600–2000)
6. Scientific name of subspecies of *Pisum sativum* referred as field peas is .... (*hortense/arvense/sinense/cogenesis*)
7. Two-third of peas production in India is from the state of .... (Bihar/Punjab/Uttar Pradesh/Assam
8. Peas require ....°C for the growth and development. (4–6/20–24/13–15/20–25)
9. Peas was originated in .... (the USA/Western China/Northern India/South Europe)
10. Dry pods of .... are liable to be shattered after maturity. (chick pea/peas/lentil/French beans)
11. Chick pea was originated in.................. (South Mexico/South Peru/South Turkey/North India)
12. .... is a major exporter of chick pea. (India/Sri Lanka/Spain/Canada)
13. Phytic acids, one of the anti-nutritional factors in pulses, can................ (reduce the digestion/act as poison to the body/reduce the availability of minerals/make the amino acids unavailable)
14. .... has weak semi-erect stem with two seeded brownish pods at maturity. (Field bean/Chick pea/Peas/Lentil)
15. Roots of .... crop have relatively poorer nodulation. (chick pea/peas/lentil/French beans)
16. .... is poor in protein, although leguminous in nature. (Field peas mature grains/Vegetable peas immature pods/Immature pods of French beans/Mature grains of French beans)
17. Scientific name of brown seeded microsperm type of chick pea is *Cicer* .... (*kabulinum/arietinum/ariticum/arietineum*)
18. The progenitor of chick peas is scientifically called *Cicer* .... (*japonicum/reticulatum/nodulosus/arieticulum*)

State TRUE or FALSE by Using 'T' or 'F'

1. Lentil is popularly grown in utera system by broadcasting in Eastern India. (T/F)
2. Asiatic type of chick pea (microsperma) has brownish smooth coated seeds with sweetish cotyledons. (T/F)

3. Garden peas are plumpy and smooth, when immature, but are wrinkled, when mature and dry. (T/F)
4. Excess nitrogen application may promote tendril formation and reduce the yields of chick pea. (T/F)
5. Most pulses require higher phosphorus application to promote the nodulation. (T/F)
6. Temperatures more than 30°C may reduce pollination and pod yield in peas. (T/F)
7. Half of Indian pulse production comes from chick pea. (T/F)
8. The duration of chick pea is extended in North India as compared to South India. (T/F)
9. Chick pea requires colder climate during growth and flowering phase than earlier stage. (T/F)
10. Chick pea is highly sensitive to water stagnation. (T/F)
11. Chick pea, also called Bengal gram is a rich source of calcium, besides protein. (T/F)
12. Chick pea is well known for its fat content among pulses. (T/F)
13. Germination of lentil requires cold temperatures of 5–15°C. (T/F)

Answers

Fill in the blanks

1. India
2. Reddish orange
3. lentil
4. Mexico
5. 1500–1600
6. arvense
7. Uttar Pradesh
8. 13–15°C
9. Southern Europe
10. lentil
11. South Turkey
12. Canada
13. reduce the availability of minerals
14. Lentil
15. French beans
16. Immature pods of French beans
17. arietinum
18. reticulatum

True or False

1. T **2.** F **3.** T **4.** F **5.** T **6.** T **7.** T **8.** T **9.** F **10.** T
11. T **12.** T **13.** T

Part 3
OIL SEEDS

Chapter 8

Rapeseed and Mustard (*Brassica spp.*)

8.1 ECONOMIC IMPORTANCE

Rapeseed and mustard are the different crops of same genus *Brassica*, studied together, used mainly for its edible oil. Rapeseed oil is obtained from the seeds of several species of *Brassica*, and the oil from different species is not distinguished in the market, since, all have similar properties.

The seeds of mustard are directly used for cooking, the rapeseed and mustard oil is used **as cooking media** throughout North India and many parts of the world. Chinese use rape-seed plants as vegetable. Traditional and other uses have been for lamp oils, soap making, high temperature and tenacious lubricating oils, and plastics manufacturing. The composition of mustard oil makes it a highly appreciated edible oil in the European Union. Rapeseed oil has also become the primary feedstock for biodiesel in Europe. Processing of rapeseed for oil production provides rapeseed animal meal as a by-product. The by-product is a high-protein animal feed. The feed is mostly employed for cattle feeding, but also for hogs and poultry (though less valuable for these).

Why the Word 'Rape' is Used in the Rapeseed?

The word is derived from Latin name ***rapa*** or ***rapum*** representing the seeds of turnip. As the mustard and other species of *Brassica* looked like turnip, the seeds of such crop were called **rapeseed**. This word has no relation, whatsoever, with the word **rape,** popularly associated with the sexual exploitation of women.

Internationally, the word **Canola** represents rapeseed and mustard, although, canola includes turnip rape (*Brassica rapa*), besides commonly considered species like *napus, juncea and campestris*. But, Canola oil represents the oil, low in anti-nutritional ingredient **erucic acid** in oil and such oil is frequently referred to as **rapeseed 00** oil or **Double zero rapeseed** oil– indicating that it has zero erucic acid.

8.2 CLASSIFICATION OF RAPESEED AND MUSTARD

Under the name rapeseed and mustard, nine important annual oilseeds belonging to the family *Brassicaceae* (*Cruciferae*) are grown in India. They are:

1. Indian mustard (*Brassica juncea* L.), commonly called rai (raya or laha), and its variation rugosa, commonly called Pahari rai
2. Indian rape toria **Lahi** (*Brassica campestris* sp. *olifera*)
3. Black mustard (*Brassica nigra*) commonly called **Banarasi rai**
4. Indian rape **Brown sarson** (*Brassica campestris*)
5. Indian rape **Yellow sarson** (*Brasssica campestris*)
6. Swede rape or **gobhi sarson** (*B. napus* L.)
7. Ethiopian mustard or **karan rai** (*Brassica carinata* A. Braun.)
8. Taramira or **tara** (*Eruca sativa* Mill), and
9. White sarson (*Brassica alba*), commonly called **ujli sarson**

On the Indian subcontinent, native species ***B. juncea* and *B. campestris* L.** and exotic ***B. napus*** L. are the important sources of edible oil than other species. Other species like *B. alba, B. nigra* and Pahari rai do not contribute to the oil production substantially, although, they are grown in India on small scale. Two exotic species are becoming popular with the farmers in the areas where winter spell is longer. The other exotic species popular in India is Ethiopian mustard (karan rai). Besides, taramira, believed to be a native of Southern Europe and North Africa is grown in the drier parts of the Northwest India.

As such, many other species of genus *Brassica* are being cultivated in other parts of the world. Some of them are *B. napus, B. rapa, B. oleracea*. **The group of rapeseed and mustard crops in other parts of the world may include the local species of *Brassica* and not necessarily the members of group relevant to India.** In general, rapeseed–mustard group of crops is largely grown under the hardy conditions of rainfed agriculture with low input management during *rabi* season, but have a good inherent potential to convert natural resources into usable biological energy.

Variation in varieties along with local names and chief characteristics are as follows:

- *Brassica campestris*, also called **Rapeseed** in English and **Sarson** in Hindi has characters, such as bold, large, round yellow/brown/dark brown/black seed which have smooth surface.
- *Brassica campestris* var. *olifera*, also called **Rapeseed** in English and **Toria** in Hindi has characters, such as spherical or ovoid shaped seed, reddish or dark brown in colour, having slightly wrinkled surface. Seeds are slightly smaller than those of Sarson (Figure 8.1).
- *Brassica juncea*, also called **Indian mustard** or **rai** or **laha** in Hindi has characters, such as small seeds, spherical or ovoid in shape having distinctly wrinkled surface. Colour of the seed is dark brown or black (Figure 8.2).

8.3 ORIGIN AND HISTORY

Mustard is cultivated in many countries of the world since olden days. *Brassica campestris* and *B. juncea* (**Toria, Rai, yellow sarson and brown sarson**) were originated in **Afghanistan**, which was a part of Northern Western areas of Indian subcontinent. It spread to other parts of the world, including China from this region.

Figure 8.1 Sarson grains.

Figure 8.2 Black Indian mustard grains.

Black mustard (*B. Nigra*) is believed to have been **originated in Eastern Europe**. *B. napus* (**Gobhi sarson**) is believed to have been **originated in Southern Europe**, *B. carinata* (**Karan rai**) was **originated in Ethiopia**, *Eruca sativa* (**Taramira**) was **originated in Southern Europe** and **North Africa**.

Historically, mentions about the use of mustard oil in Ayurveda samhita and many Vedic literature are common. It was being used as lamp oil as well as cooking media. The use of mustard oil is recorded in the last few centuries of the pre-Christian era. Its description is found in Sanskrit, Roman and Bible Scriptures and Chinese old literature.

8.4 CHARACTERISTIC FEATURES OF FEW IMPORTANT TYPES

- **Yellow and brown sarson (*Brassica campestris*):** It is widely grown in North and Central India. There are two main types—yellow and brown sarson, so named because of its seed colour. It is high yielding than toria. The crop is sown in October and Harvested in March/April after about 150 160 days. Table 8.1 shows the difference between yellow and brown sarson.

Table 8.1 Differences between Yellow and Brown Sarson

Yellow Sarson	*Brown Sarson*
Dark, glaucous, fleshy leaves.	Pale, glaucous, thin leaves.
Thick and broad pods.	Thin and narrow pods.
Dingy white or yellow, non-mucilaginous seeds.	Dark brown, brown or reddish brown, mucilaginous seeds.
Late maturing.	Early maturing.

- **Indian mustard (*Brassica juncea*):** It is very widely grown in India. The crop is sown in October/November and harvested in March/April, after about 140–160 days period. This type gives better yield than *Brassica napus.*
- **Toria/Lahi (*Brassica campestris* var. *olifera*):** It is a **short duration** crop, popularly used as catch crop or intercrop sown in September and harvested in December. It is obviously low yielder. Table 8.2 shows the difference between sarson and toria.

Table 8.2 Differences between Sarson and Toria

Sarson	*Toria*
Lower leaves and lower part of the stem are hairy with thin leaves.	Leaves and stems glabrous, leaves somewhat fleshy.
Dark coloured, mucilaginous seeds.	Light coloured, non-mucilaginous seeds.
Matures atleast 1–2 months later than toria.	Matures 1–2 months earlier than sarson.
It is sown in October/November and harvested in March/April as *rabi* crop.	It is a late *kharif* season crop, sown in September and harvested in December.
It is used as oilseed, fodder, vegetable oil and preferred for culinary purpose.	Grown for oil purpose; plants are not good for vegetable purpose.

- **Gobhi sarson (*Brassica napus*):** It is grown as an ***autumn crop***. This variety is susceptible to cold and is sown early in the middle or late September and takes about 75–100 days to mature. The variety is obviously low yielding, but it responds to irrigation and adequate fertilisation.
- **Taramira (*Eruca sativa*):** It is relatively of recent introduction into India. It is believed to be native of South Europe and North Africa. It is relatively a low yielding cruciferous oilseed crop grown in Northern India and very often grouped with rape and mustard crops. The variety is particularly adapted to poor soils and low rainfall areas

8.5 AREA AND DISTRIBUTION

Rapeseed and mustard crops are being cultivated in 53 countries spreading over the six continents across the globe covering an area of 43.5 million hectare with an average yield of 2114 kg/ha ranging from 666 (Central America) to 3716 kg/ha (Western Asia) and netted the total production of 91.87 million tons. India's contribution to the world area and production is 20.37 and 13.76% respectively. Canada (8.8 million hectare), China (7.5 million hectare) and India (8.85 million hectare) and leading producers of rapeseed and mustard in the world. Other important countries producing rapeseed and mustard are Germany (1.1 million hectare), France (1.3 million hectare) (FAOSTAT, 2023). Indian productivity stands at 1428 kg/ha.

In India, rapeseed–mustard is cultivated on 8.85 million hectare in *rabi* season (September/ October to March/April) in most parts of the North Plains. While, its cultivation in South India is almost absent, the crop distribution in different Northern states of India is as follows:

- **Indian mustard:** Assam, Bihar, Haryana, Himachal Pradesh., Jammu & Kashmir, Madhya Pradesh, North Eastern States, Odisha, Punjab, Rajasthan, Uttar Pradesh, West Bengal and Maharashtra.

- **Brown sarson:** Himachal Pradesh and Kashmir Valley.
- **Yellow sarson:** Assam, Bihar, Uttar Pradesh, West Bengal and North Eastern States particularly Meghalaya and Sikkim.
- **Toria:** Assam, Odisha, West Bengal, Meghalaya, Tripura, Haryana, Himachal Pradesh, Jammu, Madhya Pradesh, and Rajasthan.
- **Taramira:** It is confined to drought prone area of Haryana, Himachal Pradesh, Punjab and Rajasthan.
- **Gobhi sarson:** Himachal Pradesh, Haryana, Punjab and Rajasthan.

The major rapeseed–mustard growing states are Haryana, Madhya Pradesh, Rajasthan and Uttar Pradesh representing 81% of the national acreage and contributing 82.9% to the total rapeseed–mustard production. Rajasthan is the largest rapeseed–mustard growing state and alone contribute 38.2% to the production of the country from 39.3% area. The other states with substantial acreage and production are Assam, Gujarat and West Bengal.

8.6 SOIL AND CLIMATIC REQUIREMENTS

It is a prominent *rabi* crop, with particular **resistance to cold conditions**. Ideal temperature requirements for mustard and rapeseed vary from 15°C to 25°C. The germination is quick at 27°C, but for future growth needs a cool temperatures between 15°C and 20°C and at maturity, a high temperature up to 28–30°C is desirable. However, frost can damage the crop badly. At any stage of the crop, minus temperatures are harmful. Toria is especially more sensitive to cold conditions than mustard, hence, sown earlier to avoid such severity. In general, cold temperatures during night and bright sunshine during day and enough moisture ensures better yield.

The rains or cloudy weather or frost and high humidity during flowering can reduce the reproductive development and reduce the yields. Places with annual rain of 75–100 cm mostly before September are suitable for mustard cultivation. In sandy loam or sandy loam soils, one or two showers during the crop growth is necessary. During ripening phase, the climate must be dry and windless.

Mustard/rai Laha is grown in sandy to clay loam soils, with wider flexibility. But, they are best suited to alluvial soils. **Black cotton soils are not suitable to grow mustard or sarson. Terrai soils** of Uttar Pradesh, **sandy soil** of Punjab and Haryana and **heavy loam soils** of Bihar are suitable to grow yellow sarson. In general, yellow sarson require heavier soils than other rapeseeds and mustards. They cannot tolerate water logging and salinity, but soils with pH of 7–8 arc considered best for mustard and rapeseeds

8.7 BOTANICAL DESCRIPTION OF PLANT

Rape and mustard belong to family *Cruciferae* and genus *Brassica.* The botanical characters of plants of different types of rapeseeds/mustard differ substantially.

- **Sarson and toria (*B. compestris*):** It is herbaceous annual shorter than rai (mustard) between 45–150 cm (Figure 8.3). The stem is covered by waxy deposits. Leaves are sessile and glabrous or pubescent. Lower part of the leaves clasp the stalk partially or completely. Fruits are thicker than fruits of mustard and are laterally compressed with a beak one-third to half of their length. Seeds are yellow or brown with smooth seed coat.

- **Mustard (rai) (*B. juncea*):** Plants are taller (90–200 cm) erect and much branched. The leaves are not dilated at base and they do not clasp the stalk, but are broad pinnatifid. Fruits are slender (2–6 cm long) with short stout beaks. Seeds are brown or dark brown or black, with rough seed coat.

Both rapeseeds and mustard are self-pollinated, but cross pollination may occur to the tune of 10%. Flowers have 4 sepals, 4 petals of deep yellow to pale yellow colour (Figure 8.4). Ovary matures in two celled fruit (siliqua), containing seeds in two chambers separated by septum.

Figure 8.3 Toria crop in flowering.

Figure 8.4 Flowers of mustard.

8.8 VARIETIES

- **Varieties of toria:** Type 9, Bhavani, PT 30, PT 303, TH 68, Sangam, TLC 1, M-27, Aharni, BR 23
- **Varieties of brown sarson:** BSH 1, Pusa kalyani
- **Varieties of yellow sarson:** YST 151, Type 42, K 88, YS 24, Vinoy, PS 66, PSY 842
- **Varieties of rai/raya:** Sita, Bhagirathi, Varuna, Shekhar, Kranti, Krishna, Narendra rai, Vardan, Rohini, Vabhav, Gujarat mustard 1, Durgamani, Prakash, Saurabh, Laxmi, Pusa Bahar, Pusa Jai Kisan, Rajat

8.9 CROPPING SYSTEMS

Sequential cropping with *kharif* crops like cotton, paddy, corn, pearl millet, green gram is particularly more popular with toria, which is early maturing as compared to sarson and mustard group under rainfed conditions. Under irrigated conditions too, toria is used in many three crop sequences like, maize–toria–wheat; maize–toria–sugarcane; and pearl millet–toria–barley.

Intercropping sarson and mustard is more popular with number of crops. Common intercropping systems are gram + mustard (3:1), potato + mustard (3:1), lentil + mustard (5:1), wheat + mustard, barley + mustard (6:1), linseed + mustard (5:1) are commonly followed.

8.10 SEEDS AND SOWING

The land preparation includes deep ploughing, followed by two to three harrowing and final levelling. **Fine seed bed** is necessary, the crop is sown to a depth of 2–3 cm.

Planting time is a single dominant factor to decide the yield and oil content. The rate of oil development is heavily influenced by **different climatic factors**, and hence ideal time of sowing is crucial both for yield and quality. Optimum sowing time for sarson and rai is the first fortnight of October, while toria should be sown in the last week of September. Delay in sowing time should be avoided. But, across the states, some variations may occur, For example, Rai is sown in Punjab till second week of November, but in Punjab and Haryana, toria is sown in the first fortnight of September.

Seed rate varies from 5–6 kg/ha in sole crops to 1.5–2 kg/ha in intercrops. Toria is planted in rows of 30 cm, while sarson and rai are sown at a row spacing of 45 cm. In any case, thinning after 3 weeks, retaining plants at a distance of 10–15 cm in each row is essential for the successful establishment of the crop. Invariably, mustard and rapeseed are sown by drill or plough sole method using Thiram treated seeds (@2.5 g/kg).

8.11 NUTRIENT MANAGEMENT

Improved varieties of mustard and rapeseeds are particularly responsive to higher doses of fertiliser applications, as their uptake is very high, compared to many other oilseeds—both under rainfed and irrigated conditions. Applications of 80 kg N/ha under rainfed conditions and up to 120 kg/ha under irrigated conditions is found to be useful in improving the production. Similarly, a dose of 30–50 kg P_2O_5/ha and 20–40 kg K_2O/ha is found necessary to produce more yields. As sulphur deficiency is widespread, application of 40 kg sulphur/ha is found more profitable in most regions growing mustard and rapeseed. Application of nitrogen in two splits is necessary (½ at the time of sowing and remaining at the time of first irrigation or 35–40 days after sowing, whichever is earlier. Seed treatment with Azotobacter @400 g/ha is found to be useful, in addition to the above dosage.

8.12 WATER MANAGEMENT

In most situations, mustard and rapeseed crop are raised in rainless *rabi* season using stored soil moisture, and hence the success depend on the organic carbon status of the soil. Many farmers have practice to conserve the rain in bunded and levelled field or even ploughed 2–3 times during the rainy season. But the crop responds well to even 1–2 irrigations during critical stages, like pre-blooming and pod development stages. Its water requirement varies between 35–45 cm.

8.13 WEED MANAGEMENT

Weed control is crucial in the initial 30–40 days, as the growth of plants is slower in this period. Intercultivation at 20 and 40 DAS, coupled with hand weeding would be more ideal. Intercultivations can reduce the evapotranspiration losses. Some farmers even thin the crop by intercultivation operation. Most common weeds that may appear in mustard/rapeseed field include *Chenopodium album, Lathyrus spp., Melilotus indica, Cirsium arvense, Fumaria parviflora* and *Cyperus rotundus.* In severe weed infestation, pre-emergence spray of Nitrofen @1–1.5 a.i./ha or Isoproturon @1 kg a.i./ha may be able to offer effective control.

8.14 HARVESTING, POST-HARVEST AND YIELD

As soon as the **pods turn yellowish brown**, the crop should be harvested. The crop is liable to shattering and delay in harvesting may cause severe losses. Entire crop is cut at the base and stacked in threshing yard for 5–6 days. The pods easily shatter and seeds are separated by a simple beating. The seeds are separated from the wastes by slow moving air current. Nowadays, threshing is done through mechanical threshers. Cleaned seeds must be dried under the sun to bring the moisture to 8% before storage. It is possible to harvest 15–20 q/ha of rapeseeds (toria/sarson) or 20–25 q/ha of mustard/laha/rai.

8.15 PESTS AND DISEASES

- **Alternaria blight:** Caused by fungus *Alternaria brassicae*, the disease perpetuates through seed and plant residues retained in the soil. Symptoms include concentric black spots on the leaves, stem and pods. Entire pod may turn black and even rot. Control measures include spray of Mancozeb @2 g/l at an interval of 10 days immediately after symptoms are noticed.
- **Downy mildew:** Caused by fungus *Peronospora brassicae*, it is characterised by symptoms, like yellow irregular spots on the upper surface of leaves with white growth under the surface of leaves. Even inflorescence also may be affected. Resulting into its malformation or twisting and covered with white powder. The control measures include treating the seed with Apron @6 g/kg or spray of 0.2% Ridomyl or 0.1% Karthane as soon as symptoms are observed, and repeat the spray 2–3 times every 10 days.
- **Mustard saw fly:** It is most important seedling pest. The larvae of fly feed on the leaves making holes or eating away entire leaf, except midrib. In some years, it may become sporadic and the entire crop may fail. However, the pest subsides suddenly after the onset of cold season and operates mostly in the initial 30 days of the crop. Control measures include dusting 20–25 kg Methyl Parathion per hectare or spray of 0.05% Malathion will give effective control of the pest.
- **Mustard aphid/Painted bug:** It is a serious pest and main limiting factor in the production of rapeseed and mustard. Both nymphs and adults suck the sap from tender leaves, twigs, stem, inflorescence and even pods resulting into their curling and wilting of these parts. They also secrete honey dew like substances and mould will develop

on such secretions. Cold and cloudy weather aggravates the attack. Control measures include early sowing or spray of 0.05% Malathion.

- **Cabbage butter fly:** The larvae feed vigorously on the leaves, branches and pods, resulting in defoliation and even death of small plants. The control measures include spray of Monocrotophos @1 ml/l.
- **Bihar hairy caterpillar:** It is polyphagous pest and can cause severe damage. Young larvae may feed on lower epidermis of leaves, but grown up larvae feed on the entire leaf, leaving only midribs. If this attack occurs at the pod development stage, entire pods may also be eaten up or at least cause shriveling of pods. Early instars of larvae can be controlled by dusting 2% Methyl Parathion spray @20 kg/ha. Advanced instars can be controlled by the spray of Malathion @1–1.5 ml/l.

8.16 ANTI-NUTRITIONAL FACTORS

Mustard/rapeseed, its oil and its oil meal may contain anti-nutritional factors such as Goiterogens (thioglucosides or glucosinolates), tannic acid, erucic acid, sinapine (cholinester), pectins and oligosaccharides.

Why Erucic Acid is Undesirable in Canola/Rapeseed/Mustard Oil?
Erucic acid is monounsaturated omega 9 fatty acid, available in the seeds of many species of genus *Brassica*. Although, it is not proved to be dangerous to human beings, publication of animal studies with erucic acid through the 1970s, led to governments worldwide moving away from oils with high levels of erucic acid, and tolerance levels for human exposure to erucic acid have been established based on the animal studies. **Provisional Tolerable Daily Intake (PTDI)** has been fixed as 500 mg/day/person, based on the studies on myocardial lipidosis in rat. Food grade rapeseed oil is not expected to contain erucic acid more than 2%, and hence popularly called **Low Erucic Acid Rapeseed Oil (LEAR oil)**. But, Indian mustard oil contains erucic acid higher than such internationally fixed limit of erucic acid.

8.17 MEDICINAL/DIETARY BENEFITS OF MUSTARD

Mustard stimulates digestion and salivary secretion. Mustard seeds have an advantageous chemical composition, such as its protein content and fairly well balanced amino acid composition, rich in dietary fibre and natural antioxidants. In addition to its nutritional value, mustard seed flour offers rather unique functional properties. White mustard has been used effectively for food and medical applications; one of the limiting factors for human use of mustard products is the spicy flavour produced by myrosinase enzyme. Mustard seeds have high-energy content, having 28–32% oil with relatively high protein content (28–36%). The amino acid composition of mustard protein is well balanced; it is rich in essential amino acids. Mustard oil has 20–28% oleic acid, 10–12% linoleic, 9.0–9.5% linolenic acid, and 30–40% erucic acid (erucic acid is indigestible for human and animal organisms). The high erucic acid content of mustard seed could be reduced by breeding, some low erucic acid content genotypes are in cultivation in several countries. Mustard oil is rich in **tocopherols**, as a consequence of their antioxidant characteristic, they act as a preservative against rancidity.

REFERENCES

Anonymous (2010), Mustard, *ikisan,* http://www.ikisan.com/Crop%20Specific/Eng/links/up_mustardHistory.shtml.

Anonymous (2010), Report on Rapeseed, *National Multi Commodity Exchange of India*, http://www.nmce.com/files/study/rapemustard.pdf.

Anonymous (2011), Rapeseed and Mustard, *Agrisnet*, Sikkim, http://sikkimagrisne t.org/General/en/Mustard.aspx.

Anonymous (2012), Mustard and Oilseeds, in book oilseeds http://dacnet.nic.in/farmer/new/ROilseed%20Crops.pdf.

Anonymous (2012), Mustard, *ikisan*, http://www.ikisan.com/Crop%20Specific/Eng/links/up_mustardSoils%20And%20Climate.shtml.

Anonymous (2013), Canola Wikipedia, http://en.wikipedia.org/wiki/Canola. Anonymous (2013), Erucic acid Wikipedia, http://en.wikipedia.org/wiki/Erucic_acid. Anonymous (2013), Rapeseed, Wikipedia, http://en.wikipedia.org/wiki/Rapeseed.

Anonymous, 2022, Agricultural statistics at a glance, DAFW, GoI, p. 29.

FAOSTAT, 2023, https://www.fao.org/faostat/en/#data/QCL.

Kumar, Arvind, O.P. Premi and Lijo Thomas (2010), Rapeseed–Mustard Cultivation in India–An Overview, *National Research Centre on Rapeseed–Mustard*, Bharatpur, pp. 6–8. http://gcirc.org/fileadmin/documents/Bulletins/B25/B25_06Rapeseed.pdf.

Chapter 9

Sunflower (*Helianthus annuus* L.)

9.1 ECONOMIC IMPORTANCE

Sunflower was known to world prior to 19th century as **ornamental plant** with its **attractive flowers**. After recognition that its seeds have **high quality edible oil,** it is widely grown in many countries as an **oil seed crop**. Its oil is rich unsaturated **oleic and linoleic acid** (74% PUFA) and hence useful to human health. **As these fatty acids are not produced in human body, consumption of sunflower is considered as essential.** Its use is associated with healthy heart, as sunflower oil hardly allows the **blockage of blood vessels** and do not result into higher **cholesterol**. They also regulate blood pressure. The seeds contain average of **28%** oil (maximum of 36%), while kernel alone contains **44–45%** oil. It is mainly used as frying medium and for **cosmetic purposes** as **emollient**. The oil contains large quantities of vitamin E, sterols, squalene, and other aliphatic hydrocarbons, terpene and methyl ketones.

The sunflower was domesticated by USA as commercial crop, although sunflower oil was first produced by **Russian empire in 1835**. The sunflower gets its name because its flowers (inflorescence) **resemble the shape of sun**. The sunflower seeds are also used as bird feed, its meal used as **cattle feed**. Its other uses include use of kernels as food with or without roasting, making sunflower butter, sunflower **whole seed bread**, production of **non-allergic rubber,** besides used for various applications as alternative to olive oil, as some varieties of sunflower produce oil **more rich oleic acid than even olive oil**. Sunflower oil is used as body/hair oil, making dyes and different plant parts used against snake bites traditionally.

9.2 ORIGIN AND HISTORY

The wild sunflower is native to North America but its commercialization took place in **Russia**. It is commonly accepted that North American prairie states domesticated around 4400 years back (by native population), although some evidences suggest that domestication took place even in Mexico around same time. It was only recently that the sunflower plant returned to North America to become a **cultivated crop**. Even commercial extraction of edible oil from sunflower seeds was first started in Russia.

Spanish explorers took sunflower to Europe during 1500, mainly as an ornamental plant. Although, squeezing oil from sunflower seeds was patented in Britain during 1716, commercial extraction of sunflower oil became realty by 1769. Russia established large scale facility to extract sunflower oil and promoted its use in a large way during the 18th century. By the early 19th century, more than 2 million hectare of sunflower was cultivated in Russia. Russian immigrants brought cultivated form of sunflower back to the USA in 1875. It entered Canada as late as 1970. Europe got back the crop after 1970 as cultivated crop, mainly to reduce the imports of sunflower oil from Russia and the USA. Even now, Western Europe is the largest consumer of sunflower oil, but depends on its own production.

9.3 AREA AND DISTRIBUTION

Sunflower is grown over 30.14 million hectares in the world, mainly distributed in the countries like Argentina (2.4 million hectare), Romania (1.07 million hectare), Russian federation (9.9 million hectare) and Ukraine (5.2 million hectare). Other prominent countries growing sunflower are the USA and Tanzania (each with 0.7 million hectare), Turkey (0.6 million hectare) Spain (0.78 million hectare), South Africa (0.5 million hectare), Myanmar (1.8 million hectare), Kazakhstan (4000 hectare), India (0.3 million hectare), Hungary and France (each with 0.62 million hectare), China (1.00 million hectare) and Bulgaria (0.86 million hectare) (FAOSTAT, 2023).

The productivity in most European, Latin American countries is as high as 1150–2067 kg/ha, while Indian productivity is pegged at 996 kg/ha.

In India, it is mainly grown in six states like Karnataka (1.6 lakh hectare), Andhra Pradesh (0.2 lakh hectare), Maharashtra (0.3 lakh hectare) and Bihar, Odisha and Tamil Nadu (each having lesser than 20,000 hectare). In all other states, it is grown on negligible area. The average productivity is not more than 905 kg/ha, while it varies from 531 kg/ha (Maharashtra) to 1245 kg/ha (Odisha).

9.4 SOIL AND CLIMATIC REQUIREMENTS

Sunflower adjusts in climates ranging from **arid under irrigation** to **temperate rainfed** conditions, but is **highly susceptible to frost**. Mean daily temperatures for ideal growth are between 18–25°C. It is tolerant of both low and high temperatures, but more tolerant to low temperatures. Sunflower seeds can germinate at low temperatures of 4°C, but temperatures of at least 10–12°C are required for the satisfactory germination. Seeds are not affected by exposure to cold temperatures in the early germination stages. Even the seedlings in the cotyledon stage have survived temperatures down to 5°C. At later stages, freezing temperatures may injure the crop. Temperature less than 6°C are required to kill grown up sunflower plants.

Optimum temperature for later growth stages are 21–25°C, but a wider range of temperatures (17 to 33°C) show little effect on productivity. Extremely high temperatures have been shown to the lower oil percentage, seed fill and germination.

Sunflower is a **short-day plant** with a variable response to day length, but day-neutral varieties do exist. The total growing period varies from 70 days in parts of Russia to 200 days at higher altitudes in Mexico. In the subtropics, under irrigation the total growing period is about 130 days.

The crop is mainly grown under rainfed conditions on a wide range of soils. Under erratic and low rainfall, **deep soil with good water holding capacity** is required. Due to its extensive root system (0.5 to 0.7 m) soil water can be explored. Optimum soil pH is in the range of 6.0–7.5, but at lower values liming may be necessary. Sunflower is known for **tolerance to salinity** during crop establishment. However, in later growth periods sunflower is less tolerant to the salinity.

A critical time for water stress is the period, 20 days before and 20 days after flowering. If stress is likely during this period, irrigation will increase yield, oil percentage and test weight, but decrease protein percentage.

Medium to high levels of macronutrients are usually required for good plant growth. Good soil drainage is required for sunflower production.

9.5 BOTANICAL DESCRIPTION OF PLANT

Sunflower belongs to family *Asteraceae.* Although, *Helianthus* genus has many species, species *annuus* became more popular due to its oil content. Four perennial species are also identified, one of which is *Helianthus tuberoses* (Jerusalem artichoke), yielding edible tubers. Sunflower is a **tall** and **erect growing annual plant**, which flowers after reaching heights of 1.8–3 meters (6–10 ft). The present domesticated form use un-branched single headed, but earlier forms were branched and multi headed. Leaves are rough to feel, alternate, heart-shaped, and pubescent. Flowers are yellow to orange in colour (Figure 9.1) and composite with dark brown centres. The root system is well developed, but shallow (50–70 cm). The inflorescence (popularly called **flower**) is head, with characteristic peripheral ray florets with brightly coloured petals yellow, orange, as well as small incomplete flowers in the centre (called **disc florets**) which bear black seeds (botanically each seed is a fruit) (Figure 9.2). The cross-pollination is a rule, as pollen is carried by insects visiting the heads. Even the pollen from the same head does not reach the disc florets in the middle part of the disc–unless assisted by insects.

Figure 9.1 Fully grown head (inflorescence).

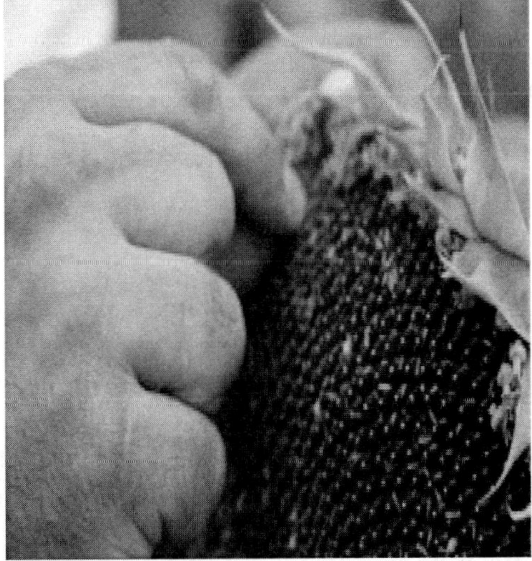

Figure 9.2 Mature seeds on sunflower head.

9.6 VARIETIES AND HYBRIDS

Varieties/hybrids released for All India cultivation

- **BSH 1:** First hybrid released during 1980, with yielding ability up to 1500 kg/ha and oil content 42%, maturing in 90 days.
- **KBSH 1:** Released from UAS, Bangalore and APSH 1 released from ANGRU, Hyderabad each with yielding ability up to 1500 kg/ha and oil content 44% maturing in 95 days.
- **MSFH 1 and MSFH 8:** Both released from Maharashtra, each with yielding ability up to 1500 kg/ha, oil content 42% and maturing in 95 days.
- **DRSF 108:** Variety released by DOR, Hyderabad during 2004; yield up to 1400 kg/ha; oil content 37%; matures in 95–100 days.
- **DRSF 113:** Variety released by DOR, Hyderabad during 2007; yield up to 1400 kg/ha; oil content 40%; matures in 90–100 days.
- **DRSH 1:** Hybrid released by DOR, Hyderabad; short; released during 2006; yield up to 1600 kg/ha; oil content 40%.

Russian varieties

Sunflower crop was introduced to India during late 1960s from Russia. Initially, many Russian varieties were being grown. Some of them are as follows:

- **Morden:** Short duration variety (matures in 80 days) with yielding ability up to 800 kg/ha and oil content 38%.
- **EC 68414:** Medium duration variety (matures in 110 days) with yielding ability up to 1000 kg/ha and oil content 42%.
- **TNAU SUF-7:** Short duration (90–95 days); yielding up to 1200 kg/ha; oil content 42%.

Varieties recommended for different states

- **Surya:** For Maharashtra; short duration variety (90–95 days) yields up to 1000 kg/ha; oil content 35%.
- **SS 46:** For Maharashtra; medium duration (100–110 days); yields up to 700 kg/ha; oil content 42%.
- **EC 68415:** For Karnataka; medium duration variety (110 days) with yielding ability up to 1000 kg/ha and oil content 42%.
- **Co 1:** For Tamil Nadu; extra early variety (65–70 days); yields up to 700 kg; oil content 39%.
- **Co 2:** For Tamil Nadu; early variety (85–90 days); yields up to 1000 kg/ha; oil content 40%. (Oil percentage refers to oil in kernel)

Hybrids recommended for different states

- **BSH 1 and KBSH 1:** Matures in 90–95 days yielding 1500 kg/ha having oil content of 42–44%.

- **LSH 1:** For Maharashtra; early variety (85–90 days); yields up to 1200 kg/ha; oil content 39%.
- **LSH 3:** For Maharashtra; medium duration (95–100 days); yield up to 1500 kg/ha; oil content 40%.
- **PSFH 67:** For Punjab; early variety (90–95 days); yields up to 1500 kg/ha; oil content 42%.

9.7 CROPPING SYSTEMS

Sunflower is a **short duration crop** in most areas and it fits into different **sequential systems** very easily in rainfed and irrigated conditions. As it can be grown in *kharif* and *rabi* season in South India, where its maximum area is concentrated, many sequential systems are adopted by the farmers. Some common sequential systems are:

Sunflower–groundnut; sunflower–black gram/horse gram; sunflower–chick pea; paddy–Sunflower; groundnut–sunflower; soybean–sunflower; and maize–sunflower.

In north India, sunflower is mainly grown in early rabi or late kharif or summer seasons.

Many replacement series of intercropping systems are also in practice, in crops offering least resistance. For example: Groundnut + sunflower (4:2), red gram + sunflower (4:2), ragi + sunflower (6:2). Sunflower + niger mixed cropping is recommended, where the seed setting is a constraint due to less movement of insects, as niger attracts more insects, helping pollination of sunflower.

9.8 SEEDS AND SOWING

Sunflower is not affected seriously by the season and day length in India. Avoiding freezing temperature, it can be sown in any month. But best results are shown in term of yields, when it is sown in the first fortnight of July, second fortnight of October, and first fortnight of March respectively for *kharif*, *rabi* and summer crops. However, in many South Indian states, where the rains start from May or June, the time of sowing will have to be preponed to June, September and December respectively. If the sowing time is altered, then the crop may be caught in rains leading to lodging, or may be exposed to cold temperatures leading to poor growth or damaged by very high temperatures reducing the pollination.

Sunflower is sown at a spacing of 60 cm between lines and 20 cm between plants at a depth not more than 3 cm. In case of extra early varieties, like Morden, the spacing will be 45 cm × 30 cm. Correspondingly, the seed requirement of all hybrids/varieties will be around 5 kg/ha, while Morden variety needs 7.5 kg/ha.

For temperate climates, the optimum planting date for early as well as late maturing varieties is between late spring and early summer. Delay in planting results in shortening of the vegetative period and early maturity, causing a decrease in head diameter and seed weight.

9.9 NUTRIENT MANAGEMENT

It is an exhaustive crop and needs high nutrition. It may extract around 175 kg N, 64–70 kg P_2O_5 and 200–225 kg K_2O per hectare, when around 14–15 q grains are harvested. This clearly indicates that sunflower require heavy fertilisation. Initial application of 8–10 tons of manure per hectare, 15–20 days earlier to the sowing, is essential for the success of the crop. Further, fertilisers supplying 60–80 kg N/ha, 60 kg P_2O_5/ha and 40–50 kg K_2O per hectare are necessary to be applied. Out of this, entire phosphorus and potassium are applied as basal dose along with two-third N and remaining dose of N is applied at the time of flowering.

Application of $ZnSO_4$ @10 kg/ha along with basal dose of fertilisers can improve the seed yield and oil yield, while foliar application of 2% Borax on heads at the time of flowering may improve the seed setting and increase the grain yield.

9.10 WATER MANAGEMENT

It is a crop that requires at least 40–45 cm water and generally grown as rainfed crop during kharif season, while it needs irrigation in *rabi* and summer seasons. Even in *kharif* season, irrigation at heading and grain filling is always advantageous. During *rabi* and summer, the irrigations will have to coincide with critical stages like germination, heading, flowering, grain development stages. Usually, a pre-sowing irrigation ensures good stand of the crop. In summer crop, it may require even 8–10 irrigations. Grain filling stage is particularly critical.

9.11 WEED MANAGEMENT

Weed free conditions will have to be maintained till 60 days of crop, as the initial stages of the crop are not able to withstand competition from weeds. Intercultivations are necessary, till knee height stage of the crop. If the weed population is very high, a pre-emergence sprays of Pendimethalin @1.5 kg a.i./ha or pre-planting application of Basalin @1 kg a.i./ha can control the weeds effectively.

9.12 AFTER CARE AND OTHER MANAGEMENT PRACTICES

Besides these practices, sunflower needs two important after care practices.

1. **Earthing up:** When the fertility status of the soil is good and/or when the rainfall is ideal, the crop grows vigorously and may suffer from lodging problems. It is advisable to earth up the crop at 30–35 days or latest after top dressing. It provides support to the stem, and grain loss could be prevented.
2. **Hand pollination:** One of the main problems faced in sunflower is improper filling of heads. Generally, the peripheral disc florets are fertilised and bear the grains. But disc florets in the centre are not properly fertilised and grain formation is poorer in the centre. This results more chaffy grains and lesser yield. To overcome this problem, hand pollination is suggested. By moving soft cloth on the head at anthesis stage it is possible to spread the pollen from ray florets to entire head, so that centrally located disc florets also get fertilised. Of course spray of Borax solution also helps in seed filling.

9.13 HARVESTING AND YIELD

Sun flower will be ready to harvest after 90–95 days in south India, while in hilly/cold areas, the crop may be ready for harvest after 140–160 days. The crop is ready for harvest, when back of the head turns yellowish brown. Differential maturity is common in sunflower. The mature heads are cut and dried under the sun to bring the moisture to 10% and later beaten or threshed by machines. Further sun drying of seeds is necessary to avoid fungal attack. A grain yield of 20 q/ha is expected from a properly managed crop.

9.14 PESTS AND DISEASES

- **Charcoal rot:** Caused by fungus *Macrophomina phaseolina*, it can damage summer crop. The affected plants mature early and have black ashy powder on the stem. The heads of such plants are under sized and yield is reduced. It can be voided by avoiding mono cropping and use of certified and treated seeds.
- **Alternaria blight:** Caused by fungus *Alternaria helianthi*, this disease is common in rainy season. Symptoms include appearance of dark brown to black round spots on leaves, which later become concentric along with chlorotic zone around them. They coalesce and result into leaf drying. The control measures include seed treatment with Apron @6 g/kg or spray of Mancozeb @1.5 g/l at least four times with an interval of 10 days.
- **Sclerotium wilt:** Caused by *Sclerotium rolfsii*, this wilt can affect the crop at any stage of the growth. Whitish growth of fungus along with mustard seed like bodies can be seen on plants near the soil surface, which later results in witting of plants. The disease can be avoided by following good crop rotation practices and keeping the field in weed free condition.
- **Cut worms:** They cut the young seedlings at the base and reduce the plant population itself to cause heavy reduction in the yield. The control measures include spray of Methyl Parathion @4 ml/l of water or dusting Methyl Parathion @25–30 kg/ha before sowing. Ridge planting may reduce the damages by cut worm.
- **Leaf sucking/eating insects:** Semi looper/jassids caterpillar/Bihar hairy cater-pillar/ tobacco caterpillar. All of them eat the foliage, partly or fully. The leaves are also attacked by aphids, jassids, which reduce the vitality of leaves, and thereby reduce the seed yields. The control measures include spray of Pyrethroid chemicals @½ ml/l water.
- **Ear head insects (Tobacco caterpillar/bugs/gram caterpillar):** They damage the flowers and eat kernels to finally reduce the yields. The control measures include dusting 5% Malathion dust in heads.

REFERENCES

Anonymous (2009), Sunflower Hybrids and Varieties, http://www.ikisan.com/crop%20specific/eng/links/ap_sunflowerSeed%20Varieties.shtml.

Anonymous (2009), Sunflower Varieties, Directorate of Oil Seed Research, Hyderabad, http://dor-icar.org.in/index.php/research/achievements/varieties-hybrids.

Anonymous (2011), An Overview of Sunflower in India, Krishisewa, http://www.krishisewa.com/cms/articles/miscellaneous/284-sunflower.html.

Anonymous (2012), FAOSTAT, http://faostat.fao.org/site/567/DesktopDefault.aspx?PageID=567#ancor.

Anonymous (2014), All About Sunflower, National Sunflower Association, http://www.sunflowernsa.com/all-about/history/.

Anonymous (2014), Crop Water Information: Sunflower, FAO http://www.fao.org/nr/water/cropinfo_sunflower.html.

Anonymous (2014), Sunflower, http://en.wikipedia.org/wiki/Sunflower.

Anonymous, 2022, Agricultural statistics at a glance, DAFW, GoI, p. 29.

FAOSTAT, 2023, https://www.fao.org/faostat/en/#data/QCL.

Putnam, D.H., Oplinger, E.S., Hicks, D.R., Durgan, B.R., Noetzel, D.M., Meronuck, R.A.. Doll, J.D. and Schulte, E.E. (1990), Sunflower, *Alternative Field Crop Manual*, Purdue University, https://www.hort.purdue.edu/newcrop/afcm/sunflower.html, 1990.

Singh, Chhidda, Prem Singh and Rajbir Singh (2003), *Modern Techniques of Raising Field Crops*, 2nd ed., Oxford & IBH, New Delhi, pp. 376–386.

Chapter 10

Safflower (*Carthamus tinctorius* L.)

10.1 ECONOMIC IMPORTANCE

Safflower was an important oilseed of the world, but lost its area and importance in many countries, including India, due to other high yielding oilseed crops like mustard, soybean, and groundnut. Historically, safflower was also associated with the production of natural dyes-again till cheaper aniline dyes were made available. Its use as a source of dyes is also limited. Nevertheless, the crop is still being grown in as many as 62 countries and well known for its high quality healthy edible oil, because of high linoleic acid (70%), along with other poly and mono unsaturated fatty acids. Safflower seeds contain 24–36% oil. Safflower occupies areas of scanty rainfall and is the source of income even in drought situations. It has number of medicinal uses, besides commercial uses of edible oil and dyes. Its green fodder, when cut young, is equivalent to that of oats/berseem in terms of nutritional value. Young shoots, used as vegetable, are rich in Vitamin A, iron, calcium and phosphorus.

Safflower oil can also be used for the manufacture of paints/varnishes, vanaspati (hydrogenated oil), cosmetics, soap and leather preservative. Oil meal from safflower is a popular cattle feed, soil amendment and is even used in the manufacture of protein enriched biscuits. In recent days, natural colouring agents are gaining more importance once again, due to health awareness about the dangers of synthetic colouring agents. Hence, safflower based colouring agents are now finding applications even in food industry.

The word ***Carthamus*** is the modification of latinised form of Arabic word called ***quartum*** or ***gurtum*** meaning colour of dyes. As safflower is grown in India since time immemorial, its Sanskrit name **kusumbha** is the basis of its present Hindi name **kusum.**

10.2 ORIGIN AND HISTORY

Confusions regarding the origin of safflower were even at the time of Vavilov, expert taxonomist. According to him, three centres of origin can be found for safflower. **India**, **Afghanistan** and **Ethiopia** were considered as the centres of origin by him due to large genetic variability and occurrence of wild forms in these countries. But, similarity of cultivated species to the naturally grown wild forms (*C. flavescens* and *C. palaestinus)* in Fertile Crescent (Turkey, Syria and Lebanon) led him to conclude that safflower was originated near East Asian countries than middle and South Eastern Asian countries. It must have been under cultivation in many parts of the world simultaneously, as many historical evidences suggest. In most of the literature, its importance as dye and as a source of oil is popularly mentioned. The way in which safflower is classified into Eastern type and Western type justifies simultaneous evolution of this crop in Fertile Crescent as well as in Indian subcontinent.

10.3 AREA AND DISTRIBUTION

Safflower is grown in Africa, Latin America, North America, Australia, and Africa besides Asian countries. Its major producers are India (1.08 lakh hectare), Kazakhstan (3.9 lakh hectare), Argentina (0.09 lakh hectare), Mexico (0.33 lakh hectare) as well as USA (0.5 lakh hectare) and China, Ethiopia, Tanzania and Turkey growing it on minor scale. The global safflower area is around 9.1 lakh hectare (FAOSTAT, 2023).

But the productivity of safflower vastly differs between countries. Turkey and USA having a productivity of 11161 kg/ha and Mexico recording the productivity as 1945 kg/ha, Indian productivity is low (826 kg/ha). Global productivity is 794 kg/ha.

In India, safflower was grown on an area of 2.32 lakh hectare during 2023, while it has been reduced to 1.1 lakh hectare during 2023. It is predominantly grown in the states of Maharashtra (27,000-hectare, accounting to 4% of the area), Karnataka (41,000-hectare, accounting for 17%) and remaining area in Andhra Pradesh.

10.4 SOIL AND CLIMATIC REQUIREMENTS

Safflower is better suited to **deep**, **fertile**, **clay rich**, but **well drained soils** with high water holding capacity–as adequate water can be stored in such soils during rainy season and crop can grow in rain free period utilising stored moisture. But instances in many parts of the world are on record to show that safflower can be successfully grown in lighter soils with poor water holding capacity—provided adequate rainfall occurs or irrigation is given during crop growth. Safflower is well known for its **tolerance to salinity** like barley, but higher salinity may cause poor germination or poor seed setting.

Safflower is a **cool season** crop, and it is grown in *rabi* season in India. Optimum temperatures for germination are about 15°C and 24–32°C for flowering. The seedlings (**rosette stage**) are tolerant to cold temperatures, while later phases need warmer temperatures. During maturity, the ideal temperatures of 30–35°C are suitable. But flowering may be affected by temperatures above 30°C. Frost at any stage is dangerous to the crop. Frost at flowering and maturity may result into sterility. The yields are lower with high rainfall (especially, beyond 750 mm annual rain), specifically at grain filling/grain maturity phases. A frost free climate for at least 120–125 days is absolutely needed. It can be grown in areas with 60–70 cm annual rainfall, although, the crop itself is more successfully grown in rainless period—if vertisols are used. In all other soils, it may need rainfall of 30–35 cm during the crop growth or arrangement for irrigation.

10.5 BOTANICAL DESCRIPTION OF PLANT

Safflower is a dwarf, bushy, profusely branching plant (Figure 10.1) belonging to family ***Asteraceae***. Each leaf has number of sharp **spines**, which are not piercing at young stages, but sharp and hard at harvest. In the initial stages, the plant looks like **rosette** (Figure 10.2) with several leaves emerging from the base of stem. Rosette stage is restricted till 30 days, beyond which quickly the branches grow from primary stem. The branching habits in safflower can be open and spreading type or narrow closed type—the expression controlled genetically and environmentally.

Each branch ends in globular inflorescence (Figure 10.3), botanically called **capitulum**, which is enclosed tightly in bracts. The capitulum has number of florets, and their opening proceeds from the peripheral florets inside centripetally over 3–5 days. The opening of flowers as well as capitulum development extends almost to 25–30 days, with the capitulum on primary branches developing faster than secondary and tertiary branches. The colour of petal varies from yellow on blooming turning to red on drying or remaining yellow even after drying. Some varieties have orange petals, which turn red on drying and even some others may have persistently white petals till maturity. The seeds are acne, shiny white, smooth with thick pericarp (Figure 10.4). They have concentrated oil in endosperm.

Figure 10.1 Branching pattern in safflower.

Figure 10.2 Rosette stage of safflower.

Figure 10.3 Flowering in safflower.

Figure 10.4 White safflower seeds.

Safflower has deep **strongly developed tap root system**, sometimes main anchorage root going to a depth of even 2–2.5 m. But, effective rooting zone extends generally to 1–1.5 m, offering it an excellent capacity to withstand the drought.

10.6 VARIETIES/HYBRIDS

Many traditional varieties like Manjira (Andhra Pradesh), Nag-7, Tara and N06-2-8 (Maharashtra), A-1 and A-300 (Karnataka) are popularly grown by the farmers even now. But Directorate of Oilseed Research released a safflower variety called DSH 129, with high yield potentiality up to 2200 kg/ha with an oil content of 30%.

10.7 CROPPING SYSTEMS

Safflower is mostly grown as intercrop with winter crops like wheat, barley, chick pea, *rabi* sorghum and coriander, in various row proportions. Most popular row proportions are 3–5 rows of safflower with 9–12 rows of other *rabi* crops. Some farmers grow safflower as sole crop, when it is rotated with *kharif* crops like groundnut, cotton and green gram.

10.8 PESTS AND DISEASES

- **Safflower aphid:** Yields are seriously affected, when aphid attack is severe. They suck the sap and **reduce the vitality of plant** and reduce the photosynthetic ability of plants. It can be controlled by the spray of Metasystox or Dimethoate @1 ml/l.
- **Safflower caterpillar:** They **defoliate the leaves** and **damage tender capsules**. It can be controlled by dusting 2% Methyl Parathion @20–25 kg/ha or spray of Methyl Parathion solution @2 ml/l.
- **Safflower stem fly:** Maggots bore into stems resulting in **swellings**. Plants wilt droop and die often resulting in drying of tips or even loss of the entire plant. It can be controlled by the removal of affected plants in the initial stages.
- **Rust:** Caused by the fungus *Puccinia carthami*, rust is identified by slightly discolouration of cotyledons accompanied by **drooping** and **wilting of seedlings**. Attack at later stages, may spread to entire plant, where small round pustules are grown on stem, and leaf. The disease is controlled by treating the seeds with Thiram or Bavistin @3 g/kg seed.
- **Cercospora leaf spot:** It can attack all parts of the plant and is identified by circular brown spots on lower leaves in the initial stages. The leaves turn yellow and get **distorted**. It can be controlled by the spray of Zineb or Mancozeb @2 g/l of water.

10.9 PRODUCTION TECHNOLOGIES

- **Land preparation:** Sole crop of safflower needs **deep ploughing** with MB plough followed by three harrowing and planking. As it is grown using stored moisture, care must be taken to ensure that soil sufficient moisture is available at the time sowing, even after land preparation.

- **Seeds and sowing:** Healthy seeds of safflower treated with Thiram or Bavistin @3 g/kg of seed should be used. Nearly, 15–20 kg seeds are needed per hectare for the sole crop, while, for the intercrop, it can vary from 4–12 kg depending on row proportions. Sowing is usually taken up in the first week of October to first week of November. Delayed sowing may reduce the yields, especially in the hot regions if the flowering coincides with high temperature beyond 30°C. Invariably, the crop is sown by drill at inter row spacing of 45 cm, placing the seeds deeper at 5–6 cm.
- **Manures and fertilisers:** Application of manure before final preparation of land @5–8 t/ha (heavy soils) or 12–15 t/ha (sandy and light soils) is necessary for the sole crop. For intercrop, no separate dose of manure is given, when the same is supplied to the main crop. A dose of fertilisers @40:40:20 kg N, P_2O_5 and K_2O per hectare is necessary for the sole crop. For intercrop, one-fourth to one-third dose is necessary to be supplemented with the fertilisers given to the main crop.
- **Water management:** As the crop roots are able to extract the soil moisture from deeper layers (even up to 1–1.5 m) in majority of cases, it is grown as un-irrigated crop. In case of sandy and lighter soils, it may be advantageous to irrigate once after 30–35 days. Its water requirement is 30–35 cm, and hence can be grown under limited soil moisture.
- **Weed control:** Critical period for competition by weeds extends up to 40 days, till which stem elongation and branching do not happen. Control of the weeds till such time by hoeing after 20 and 40 days of sowing will ensure good control of the weeds.
- **Topping:** When central shoot bears flowering head, the crop is topped (removal of apical part of the central shoot), generally, after 35–40 days, to promote the branching and to ensure higher yields.
- **Harvesting and yield:** The crop matures in 120–140 days, which is indicated by the yellow colour developed by the stem and leaves. Harvesting should always be done in the morning hours to avoid the trouble caused by the spines on the leaves (in mornings they are little softer). Entire plant is pulled and stacked for 6–9 days and later threshed and winnowed. The yield from sole crop can be as high as 15–20 quintals/ha (irrigated) or 10–12 q/ha (un irrigated), if the crop is well maintained. Intercrop yield varies from 4–8 q/ha.

10.10 ANTI-NUTRITIONAL FACTORS

Safflower oil cake (called **oilseed meal**) is commonly fed to different types of animals. Such meal may contain **matairesinol** (giving bitter flavour to the feed) and 2-**hydroxyarctiin** (which have cathartic properties). They may affect the health of cattle in many ways including digestion.

REFERENCES

Anonymous (2009), *Production Trends of Safflower*, http://oilseeds.dacnet.nic.in/oilseeds/safflower/safflowerproductiontrends.html.

Anonymous, 2022, Agricultural statistics at a glance, DAFW, GoI, p. 29.

Blain, Robert (2007), *Nutrition and Feeding Organic Pigs*, CAB International, p. 100, http://www.nariphaltan.org/nari/pdf_files/safflowerbook.pdf.

FAOSTAT, 2023, https://www.fao.org/faostat/en/#data/QCL.

Singh, Chhidda, Prem Singh and Rajbir Singh (2003), *Modern Techniques of Raising Field Crops,* 2nd ed., Oxford and IBH, New Delhi, pp. 367–375.

Chapter 11

Linseed (*Linum usitatissimum* L.)

11.1 ECONOMIC IMPORTANCE

Linseed or flax is an important oil cum fibre crop, grown in many cooler parts of the world, as a commercial crop. Its species name ***usitatissimum*** was derived from Greek word, meaning **most useful** and genus name *linum,* meaning **fibre**, popularly called **linen**. It is used as fibre crop to manufacture linen clothes. Till nineteenth century, linen clothes were commonly used, when cotton clothes over took them. Now, linen is used to manufacture specific fire resistant clothes. Flax fibres, which are extracted from stem (called **bast fibres**), are three times stronger than cotton fibres. Flax fibres are also useful to make ropes, canvas, and high quality papers (used as paper for currency notes). Conventionally, crop is called flax, when produced for fibres soil and as linseed, when grown for oil.

Flax/linseed seeds are consumed directly for many culinary purposes and its edible oil is used for various products. More popularly, its oil is used in the preparations of paints and varnishes as drying oil, printing inks and in the manufacture of linoleum. In most Western literature, it is called **flax**. The seeds have 35–45% oil contents, 28% fibre and 20% protein. Flax seeds are consumed directly, roasted, or powdered before consumption or even sprouted and consumed or seeds are crushed and linseed oil is extracted and used for many culinary and bakery applications. Linseed/flax oil **is remarkably resistant to rancidity** (oxidation of oil by the exposure to air) and is stable under room temperatures. Linseed oil is rich in unsaturated fatty acid, linolenic acid (51%) and small content of saturated fatty acids, like palmitic acid (7%) and stearic acid (3–4%). The linseed/flax oil is not used as cooking medium like other vegetable oils.

Flax seeds have certain medicinal uses also. Its seeds contain high levels of *dietary fibre* as well as *lignans*, an abundance of *micronutrients* and omega–3 fatty acids. Consumption of flax seeds may lower cholesterol levels. Flax seeds may also lessen the severity of diabetes by stabilising the blood sugar levels. There is some support for the use of flax seed as a *laxative* due to its dietary fiber content, though excessive consumption without liquid can result in intestinal blockage. Consuming large amounts of flax seed may impair the effectiveness of certain oral medications, due to its fibre content, and may have adverse effects due to its content of *neurotoxic* cyanogen glycosides and immunosuppressive cyclic nonapeptides. One of the main components of the flax is *lignan*, which has plant estrogen as well as *antioxidants* (flax contains up to 800 times more lignans than other plant foods contain).

11.2 ORIGIN AND HISTORY

The earliest archaeological evidence indicates that it was known in Europe and the near East, around 8000 BC. It is thought, therefore, that flax cultivation began around 6000 BC. It has been accepted that linseed was originated in the **Eastern part of the Mediterranean sea**, in the regions of **Fertile Crescent** (countries Turkey, Syria and Lebanon) which later spread towards India. Flax was an important crop in Neolithic Europe—it was found in Southeast Europe dated about 3500 BC.

Records from Babylonia and Anatolia from 3000 BC show that flax was regularly cultivated in the ancient Egypt for its seeds, for linseed oil, and for fibres to make linen. But, the Greeks cultivated the plant for getting the oil for the lamps.

Throughout the Middle Ages flax was cultivated in many European countries. Flax remained the most important vegetable fibre throughout the Europe and the Near East, until cotton became more easily and cheaply obtained after 1800. Before 1800, Germany and Russia were the main flax growers and exporters. Before cotton cultivation became widespread, there was a thriving linen industry in the USA.

11.3 AREA AND DISTRIBUTION

At global level, linseed is a **minor crop** both in respect of its oil and its fibre, as compared to other major oil seed crops and fibre crops. It is grown over an area of 3.08 million hectare with a productivity of 843 kg/ha. It is mainly grown in the countries like Canada (2.38 lakh hectare), China (2.01 lakh hectare), Ethiopia (0.82 lakh hectare), India (2.40 lakh hectare), Kazakhstan (7.26 lakh hectare), Russia (12.90 lakh hectare) and the USA (0.64 lakh hectare).

Even at Indian level, linseed is a minor oil seed crop, as its share in total area under oil seeds is as less as 3%, while it contributes 1% of national oil seed production. It is grown over an area of 2.38 lakh hectare, distributed mainly in the states of Madhya Pradesh (0.84 lakh hectare), Uttar Pradesh (0.34 lakh hectare), and Jharkhand (0.53 lakh hectare). Other states, where linseed is grown on limited scale are Karnataka, Chhattisgarh and West Bengal. The productivity of linseed is also very poor (698 kg/ha) and in the states of Rajasthan (1273 kg/ha), Bihar (847 kg/ha) and Madhya Pradesh (729 kg/ha), the productivity is equal or above national average, while in other states, it is very poor.

11.4 SOIL AND CLIMATIC REQUIREMENTS

Linseed is a **cool season** crop, grown in the latitudinal range 10° and 65°N or S. Its cultivation is confined to low elevations, but it can be successfully grown up to 770 metres. Areas with the annual rainfall from 45–75 cm are best suited for its cultivation. The seed-type linseed does well under moderate cold, but the fibre crop grows best in cool moist climates of Europe and Canada. In India, the crop is grown in the *rabi* season from September–October to February–March. The temperature during the vegetative development of the crop should be cool. High temperature above 32°C reduces the yield, the oil content of the seed and the quality of the oil.

Moderate temperatures (21–26°C) are ideal. The plants are susceptible to frost. Frost can damage the crop at flowering. Most of the crop grown in India is raised after the rainy season is over, the water requirement of the crop being met from the moisture stored up in the soil. It is also grown under irrigation in dry climate. It is a long day in photoperiodic response.

Linseed can be grown in almost **all types of soils** where sufficient moisture is available, but it will do better on **heavier soils** having more water retention capacity. It is tolerant to fairly wide ranges of pH values. It grows best on well drained loam to clayey loam soils rich in humus. In Madhya Pradesh and Maharashtra, linseed is grown largely on the black cotton soils having high clay and lime content. It is also grown on the light alluvial soils of Uttar Pradesh, Bihar and West Bengal.

11.5 BOTANICAL DESCRIPTION OF PLANT

Linseed belongs to the family *Lineacea*, and grows as perennial or annual, depending on the varieties, conditions and types of cultivation. It is an erect, herbaceous annual with branches as corymb.

Two types of *L. usitatissimum* are cultivated—**linseed type**, grown for oil extracted from the seed, is a relatively short plant which produces many secondary branches, and **flax type,** grown for the fibre extracted from the stem, which is taller and is less branched.

L. usitatissimum has a **short tap root with fibrous branches**, which may extend 90–120 cm in light soils. Leaves are simple, sessile, linear–lanceolate with entire margins, and are borne on stems and branches. The inflorescence is a loose terminal **raceme** or **cyme**. Flowers borne on long erect pedicels (Figure 11.1) are hermaphrodite, pentasepalous, pentapetalous (petal colour blue) five stamens, and a compound pentacarpellary pistil, each separated by a false septum. The fruit is a capsule, composed up to 10 seeds. The seed is oval, lenticular, 4–6 mm long with a smooth, shiny surface, brown to light–brown in colour (Figure 11.2). Seeds contain 35–45% oil content and 20–25% protein.

Figure 11.1 Flowering in linseed.

Figure 11.2 Smooth, shiny flat seeds.

11.6 VARIETIES

Different varieties are developed for different states, depending on the localised agro–climatic situations. Important varieties are shown in Table 11.1.

Table 11.1 State-wise Recommended Important Varieties

State	*Recommended Varieties*
Bihar	T-397, Garima, Sweta, Subhra, Gaurav, Nagarkot
Haryana	K 2, LC 54, Himalini, Pusa 2, Pusav 3, Nagarkot
Himachal Pradesh	K 2, LC 185, LC 54, Himalini, Pusa 2, Pusa 3, Janki, Surbhi, Jeevan, Nagarkot
Punjab	K 2, LC 185, LC 54, Himalini, Pusa 2, Pusa 3 and Nagarkot
Rajasthan	LC 54, Himalini, T 397, Jawahar 23, Chambal, Pusa 2, Pusa 3, Kiran, Jeevan, Gaurav
Madhya Pradesh	Jawahar 1, Jawahar 7, Jawahar 17, Jawahar 552, Sheetal
Uttar Pradesh	T 397, Neelam, Hira, Jawahar 23, Pusa 2, Garima, Sweta, Subhra, Kiran, Sheetal, Gaurav, Padmini
Maharashtra	Jawahar 23, C 249, Pusa 2, Kiran, Sheetal
Odisha	Jawahar 23
West Bengal	B-67

11.7 CROPPING SYSTEMS

Linseed is grown in rotation with many *kharif* crops like sorghum, pearl millet, groundnut, cowpea and soybean. It is also intercropped with the crops like barley, wheat, gram and mustard. More popularly, in the Eastern part of India, it is grown as **relay crop**, where the linseed is grown by broadcasting the seeds in between standing crop of paddy as **praire** crop.

11.8 PRODUCTION PRACTICES

- **Land preparation:** One deep ploughing with MB plough followed by 2–3 harrowing and final levelling ensuring fine seed bed is required for sowing linseed. Of course, before final levelling, 4–5 tons manure/hectare needs to be applied.
- **Seeds and sowing:** Ideal time to sow the crops is between the first week of October and the first week of November in North Indian situations. In 'uteri' in Madhya Pradesh as well as 'pair' systems, linseed is sown in standing crop of paddy. In such situations, the sowing is preponed by one month. The objective to pre pone the sowing date is to make use of soil moisture more effectively. Sowing is invariably done by using seed drill or plough sole method, using 20–30 kg seeds/hectare for sole crop and 8–10 kg seeds/hectare in intercrops. However, for utera or praire crop, 10 kg seeds are broadcasted.
- **Manures and fertilisers:** Application of manure 15 days in advance of sowing is essential. Application of fertilisers supplying 50 kg N, 40 kg P_2O_5 and 40 kg K_2O per hectare is sufficient. Entire dose of fertilisers is applied at the time of sowing.
- **Water management:** Mostly, the crop is grown as unirrigated crop. But the crop responds well to 1–2 irrigations (after 30–40 days and second before flowering).

- **Weed management:** As linseed is poor competitor, the crop is likely to be lost, if inter-cultivation is not done after 20–25 days as well as 40–45 days of sowing. Thinning operation after 2 weeks is also essential to retain seedlings every 10–15 cm. If necessary, one hand weeding may be provided. Generally, herbicides are not used, looking to its low yield potential.
- **Plant protection:** The crop may be affected by diseases like rust, wilt, powdery mildew and leaf spot. Spray of Mancozeb @2 g/l or dusting with 20 kg sulphur/ha or spray of wettable sulphur @3 g/l will be able to control the diseases. The crop may also be affected by insects like midge, leaf miner, and cut worm. The crop need protection from them by the spray of Metasystox or Dimethoate @1 ml/l as well as the spray of Dichlorophos @ one-half ml/l or dusting Methyl Parathion @20–25 kg/ha.
- **Harvesting and yield:** The crop takes around 130 –140 days to mature. The indication of maturity is provided by yellowing of stem and withering of leaves. Entire plant is cut and bundled and transferred to threshing yard, where they are beaten by sticks and the seeds are separated. A sole crop of linseed may give a yield of 15–20 q/ha under ideal management, while an intercrop give a yield of 5–10 q/ha.

11.9 ANTI-NUTRITIONAL FACTORS

The seeds of linseed and its oil can contain anti-nutritional factors such as pancreatic and salivary amylase inhibitors, trypsin and chymotrypsin inhibitors and lectins. The oil meal may also have neurotoxic cyanogen glycosides and immunosuppressive cyclic nonpeptides. Large quantity of linseed meal consumption may invite many health problems. However, different methods of extrusion cooking may reduce these factors in seeds.

REFERENCES

Anonymous (2005), *Production Trends of Linseed*, http://oilseeds.dacnet.nic.in/oilseeds/linseed/linseedproductiontrends.htm.

Anonymous (2010), *Biology of Linum usitatissumum*, Canadian Food Inspection Agency, http://www.inspection.gc.ca/plants/plants–with–novel–traits/applicants/directive–94–08/biology–documents/linum–usitatissimum–l–/eng/1330979709525/1330979779866#B2.

Anonymous (2011), Natural History Museum, http://www.nhm.ac.uk/nature–online/life/plants–fungi/seeds–of–trade/page.dsml?section=regions&ref=flax&cat_ref=®ion_ID=4& time_ref=&page=origins&origTimeID=&origTimePoint=&origTpTitle=&origPage=.

Anonymous (2012), FAOSTAT, http://faostat.fao.org/site/567/DesktopDefault.aspx?PageID=567#ancor.

Anonymous, 2022, Agricultural statistics at a glance, DAFW, GoI, p. 29.

FAOSTAT, 2023, https://www.fao.org/faostat/en/#data/QCL.

MODEL QUESTIONS AND KEY ANSWERS FOR RABI OIL SEEDS

Choose Most Appropriate Answer and Fill in the Blanks

1. Rapeseed and mustard seeds have oil percentage of .... (46–48/25–28/36–42/40–45)
2. *Toria* has .... duration and hence suitable for many sequential systems (very short/short/medium/long)
3. .... acid in mustard oil is considered as anti-nutritional factor (erucic/oxalic/brassic/inamic)
4. Mustard is .... pollinated crop (solely cross/often cross/solely self/often cross)
5. The ideal temperatures required at maturity of mustard/rapeseed is .... °C (15–20/30–35/28–30/20–24)
6. Ideally, soil pH to grow mustard/rapeseed is .... (6–7/7–8/8.5–9/5–7)
7. .... country has maximum mustard/rapeseed area (USA/Canada/India/China)
8. *Brassica nigra* (also called Banarasi rai) was originated in .... (Banaras/Ethiopia/Eastern Europe/Brazil)
9. Mustard/rapeseed requires .... element in large quantity from among secondary or micronutrients (calcium/sulphur/zinc/mangenese)
10. .... nature of pods, requires mustard/rapeseed to be harvested on time (dehiscent/hard/bristle/succulent)
11. Scientific name of linseed is *Linum* .... (*usitasimum/usitatum/usitatisimum/usitatisinum*)
12. Linseed fibres are .... and .... than sunhemp fibres (weaker and softer/weaker and harder/stronger and softer/stronger and harder)
13. Flax seeds have .... oil (24–36%/45–48%/35–45%/52–55%)
14. .... country has maximum area under linseed (China/India/Canada/Russia)
15. Ideal temperatures to grow linseed is in the range of ....°C (15–19/21–26/25–30/12–20)
16. .... state has 40% of Indian area under linseed (Haryana/U.P./M.P./Orissa)
17. Dye is prepared out of flowers of .... oil seed crop (sunflower/linseed/castor/safflower)
18. Safflower oil is known as healthy, because it contains .... acid (linoleic/oleic/stearic/myristic)
19. Latest global data shows that................has maximum area under safflower (India/USA/Russia/Kazakisthan)
20. Safflower belongs to family .... (Compositae/Aateraceae/Teliaceae/Malvaceae)
21. .... seeds contain high dietary fibre and lignans and hence recommended to reduce blood colesterol in human beings (Safflower/sunflower/linseed/groundnut)
22. .... crop is considered as most drought resistant crop among oilseeds (safflower/sunflower/castor/linseed)

23. .... crop among oil seeds is highly responsive to Boron oilseeds (safflower/sunflower/castor/linseed)
24. .... oil is imported by India in large quantity (sunflower/palm/coconut/soya)
25. The acreage of sunflower is highest in .... country (USA/India/China/Russia)
26. Linseed belongs to family (*Lineaceae*/*Euphorbiceae*/*Compositae*/*Asteraceae*)
27. DOR, Hyderabad mainly deals with research on .... (edible oils/all oils/all oilseeds/rice)
28. Morden variety of sunflower takes .... days for maturity (80/95/110/125)
29. Germination phase of sunflower requires at least .... °C temperature (20–24/10–12/15–18/25–30)
30. Artificial hand pollination is necessary for .... (oil seed crop) (linseed/safflower/sunflower/soybean)
31. Which oil is extraordinarily resistant to rancidity (safflower/sunflower/linseed/olive)
32. Linseed is rich in .... fatty acids (oleic/linoleic/linolenic/myristic)

State TRUE or FALSE by Using 'T' or 'F'

1. Linseed was originated in China. (T/F)
2. Area under safflower is diminishing in India and the world. (T/F)
3. Most consumed edible oil in India is mustard oil. (T/F)
4. Safflower and sunflower belong to the same family. (T/F)
5. The fibre is extracted from the roots of flax plant. (T/F)
6. Maharashtra has the maximum acreage of safflower. (T/F)
7. BSH 1 is an important hybrid of mustard. (T/F)
8. Two types of sarson, namely yellow and red sarson are commonly cultivated. (T/F)
9. Mustard has higher yield potentiality than sarson. (T/F)
10. Safflower sends its roots laterally and horizontally. (T/F)
11. Linseed oil is commonly used in the manufacture of paints, as it dries quickly. (T/F)
12. In many advanced countries, mustard oil is converted into diesel. (T/F)
13. The maximum yield potentiality of sunflower is as high as 30–35 q/ha. (T/F)
14. More branching is stimulated by topping in linseed. (T/F)
15. Earthing up operation is necessary in sunflower to reduce insect attack. (T/F)
16. Safflower can tolerate the cold climates during germination. (T/F)
17. Nitrogen can be used as pre-emergence herbicide in mustard and rape seeds. (T/F)
18. *Niger* is grown along with safflower to increase the productivity of safflower. (T/F)
19. Linseed oil is not edible, but its seeds are edible. (T/F)
20. Oil from mustard and rapeseed has anti-nutritional factor called palmitic acid. (T/F)

Answers

Fill in the blanks

1. 36–42%
2. Short
3. Euric
4. often self
5. 28–30°C
6. 7–8
7. Canada
8. Eastern Europe
9. sulphur
10. Dehiscent
11. Linum usitatissimum
12. Stronger and softer
13. 35–45%
14. India
15. 21–26°C
16. Madhya Pradesh
17. Safflower
18. Linoleic acid
19. Kazakhstan
20. Asteraceae
21. Linseed
22. Safflower
23. Sunflower
24. Palm
25. Russia
26. Lineaceae
27. All oilseeds
28. 80
29. 10–12°C
30. Sunflower
31. Linseed
32. Linoleneic

True or False

1. F	**2.** T	**3.** F	**4.** T	**5.** F	**6.** T	**7.** F	**8.** F	**9.** T	**10.** T
11. T	**12.** T	**13.** F	**14.** F	**15.** F	**16.** F	**17.** T	**18.** F	**19.** T	**20.** F

Part 4
SUGAR CROPS

Chapter 12

Sugarcane (*Saccharum officinarum* L.)

12.1 ECONOMIC IMPORTANCE

Sugarcane is an important commercial crop of India, influencing nation's economy in many ways. Besides directly supporting 5–6 million farmers by being a cash earning crop, sugarcane supports large number of industries like sugar mills (more than 527) producing refined sugars, distilleries (more than 250), producing liquor grade ethanol, millions of jaggery manufacturing units (mostly handled by farmers themselves), number of chemical industries (for example: citric acid) and few cooperative bora sugar units (unrefined brown powdered sugar), and millions of families dependent on them. In urban centres, freshly extracted sugarcane juice is also being widely sold as sweet drink. Sugar and jaggery are sweeteners, used in daily life of almost entire population for various culinary, confectionary preparations and sweet meats. The sugar exports also add to the expanding sector of agricultural exports.

Besides these contributions to the economy, sugar industry produces huge quantity of molasses (800 million l/annum), which is a source of producing ethanol, which can replace petrol partially and reduce dependence on the imports of petroleum products. The sugar mills produce large quantity of power (under co-generation) by using massive quantity of bagasse (10 –12 million tons annually) and even sell the power to the national/state owned power distribution grids. Sugar mills also produce large quantity of biodegradable press mud (6 million tons/annum), which is conveniently used as manure.

Many other tropical countries known for intensive cultivation grow sugarcane and their national economy is heavily influenced by it.

12.2 ORIGIN AND HISTORY

The cultivated ***Saccharum officinarum*** (also called **noble cane**) was evolved by repeated back crossing of *S. officinarum* of New Guinea with wild *S. spontaneum* to improve the commercially important characters, like high sucrose, low fibres, long inter nodes, late maturity, long leaves, etc. The attainment of such characters by natural crossing is called **nobilisation**.

Sugarcane was cultivated in India since Vedic period and sugar/jaggery were known to early Indians. The word **sugar** was generated from the Sanskrit word, called **sharkara**. Earliest recoded writing on sugarcane was found in Indian literature of 1400–1100 BC. However, in those days, *S. barberi* must have been cultivated. About 6000 years back, sugarcane must have been domesticated in New Guinea and must have travelled to India through Indo–Myanmar region and later led to the nobilisation process. But early growers of sugarcane must have tasted it only for sweet juice. It was in India, crude forms of sugar manufacturing started. In the eighth century, Arab travellers carried sugarcane to the Mesopotamia, from where it must have spread to the Mediterranean regions and other parts of Africa. British settlers in South Africa developed sugarcane cultivation using African serfs during seventeenth century. Brazil and other Central American and Carribean islands got sugarcane late in the eighteenth century.

In the historical records, sugarcane and sugar is associated with number of events like enslaving Africans and Indians for sugarcane cultivation, trading of sugarcane lands by France for a portion of Canada, Cuban sugar exported to USSR to regulate domestic prices, etc.

12.3 CLASSIFICATION AND ORIGIN

Although, genus *Saccharum* has many species, three cultivated species of sugarcane are economically important. They are:

- ***Saccharum barberi*:** Originated in **Northern India**, characterised by short, thin fibrous stalks, narrow leaves, short inter nodes, low to medium in sucrose and early maturity.
- ***Saccharum sinense*:** Originated in **China**, characterised by long thin more fibrous stalks, broad leaves, low to medium in sucrose, long zig-zag inter nodes and early maturity.
- ***Saccharum officinarum*:** Originated in **New Guinea**, characterised by thick juicy canes, rich in sugar low in fibres, long inter nodes and late maturity.

12.4 AREA AND DISTRIBUTION

Sugarcane is grown in 101 countries, and is distributed between 36°N and 31°S latitudes. Globally, sugarcane is grown over 27 million hectare land with a production of 2025 million tons (FAOSTAT, 2023) at mean productivity of 74 t/ha. Global area is mainly distributed in Brazil (10.0 million hectare), India (5.8 million hectare), China (1.3 million hectare), Pakistan (1.1 million hectare), Thailand (1.6 million hectare). All other countries have sugarcane area less than 1 million hectare. In many countries, sugarcane productivity is as high as 121t/ha (Peru) and 116 t/ha (Guatemala). At global level, sugar is not only produced from cane, but also from sugar beet. But in tropical sugarcane growing areas, sugar is invariably produced from cane.

In India, sugarcane area is distributed in the states of Uttar Pradesh (2.18 million hectare) Maharashtra (1.23 million hectare), Karnataka (0.64 million hectare), Gujarat (0.22 million hectare), Tamil Nadu (0.15 million hectare) and Bihar (0.21 million hectare), leading in its area. Other states have lesser area. Highest state level productivity of 105 t/ha is recorded by Tamil Nadu and minimum productivity is recorded by Bihar (57 t/ha). Sugarcane productivity is higher in **lower latitudes** (Southern states) than **higher latitudes** (Northern states). Among tropical South Indian states, Maharashtra leads in area, while productivity is higher in Tamil Nadu and

Karnataka. Nevertheless, maximum area is found in Uttar Pradesh supported by maximum number of sugar mills.

Around 45% of sugarcane is used to produce jaggery (a solidified sweet block made out of boiling sugarcane juice) and remaining area supplies the cane to the sugar mills. Jaggery production takes place at individual farmers' level using indigenous technologies. Both jaggery and sugar are in persistent demand in India throughout the year.

12.5 TRENDS IN SUGAR PRODUCTION

Sugarcane is conventionally grown in our country to manufacture sugar by sugar mills or to manufacture jaggery by farmers. A sugarcane grower should be able to sell his cane to the sugar mills or prepare jaggery and he has no option either to keep the cane for long time (as it gets dried up within 3–5 days of harvest) or to sell in any other open market. Hence, sugarcane production is completely dependent on sugar mills or jaggery units to realise the profits. The sugar mills, are in dictating terms on farming community, as far as the time and quantity of cane procurement is concerned. Expansion of sugarcane area invariably depends on establishment of sugar industry, as jaggery manufacturing by farmer himself needs a huge investment and large labour force. With no specific control over expansion of the sugarcane area (as no planned cropping takes place) by any authority, the farmers are always vulnerable to either delayed procurement by sugar mills or delayed payment of proceeds by such mills.

On the other hand, sugar mills have their own set of problems. A sugar mill needs huge investment. A part of sugar manufactured by a mill has to be kept for distribution on subsidised rates under government sponsored **Public Distributions System (PDS)** under Central Act. Remaining part of the sugar is allowed to be sold in the open market. Hence, the sugar mills have to depend on payment to be received from Government to meet the manufacturing expenses. In addition, a sugar mill runs for not more than 250 days in a year, even with best cane procurement. The idle period is generally used to maintain the machines. But, the overhead costs, interest pile up even in off season of around 110–120 days. The sugar mill will have to run the show and pay the farmers amidst all these problems. Cogeneration of power, selling press mud as manure and selling molasses or grooming distillery to use molasses are some options to assure themselves a good profit.

The latest trend in sugarcane cultivation is to grow it as energy crop to harness its potentiality to manufacture ethyl alcohol, a popular replacement of petrol. Hence, sugarcane can also be called **biofuel crop**.

12.6 SUGAR MILLS AND CANE PROCUREMENT

India has 527 sugar mills, with a combined crushing capacity of 24 million tons per day. Indian sugar mills invariably have recorded a recovery of 9–11% sugar. The demand for it is around the year and by all sections of the population. The cane procurement should be as extended as possible to run the factory for a longer period, obviously not more than 250 days. The vexed problem to sugar industry is to arrange the cane supplies in tune with daily crushing capacity which is generally against natural phenomenon of crop maturity in a short period of 1–2 months.

Over-aged or under-aged canes result into **low recovery**. But, mill cannot handle more than its installed crushing capacity even in peak season. Such situations invariably results into long queues of sugarcane laden vehicles waiting to be unloaded in sugar mills and handling dried canes due to drying of canes before crushing. To avoid this, a sugar mill adopts strategies like **issue of permits** to ensure cutting of the canes as per the schedule of crushing or even arranging the cutting of the crop and transporting the cane by the mill staff, etc. Some sugar mills even **supply the seed canes** to make sure that farmers plant the canes to suit to crushing schedules. To make sure that mills do not run short supply of canes, sugar mills ensure that revenue authorities issue orders to prohibit the movement of canes outside the jurisdiction of a mill or even prohibit the farmers to make jaggery in factory jurisdiction area.

One more issue related to sugar mills is about recovery. The recovery percentage will decide the sugar output per tons of cane. But, this depends on the age of harvested cane, time taken to transport it to the mill, irrigation facility, fertilisers applied and more than all these, variety of cane grown. While the recovery percent varies during total length of the crushing period, sugar mill always aims to achieve maximum recovery. In order to boost up the recovery, many mills promote the use of sugar rich cane varieties and arrange to distribute the seed canes of such varieties. But unfortunately, most sugar rich cane varieties are poor yielders and farmers are not inclined to grow such varieties, as they are paid on the basis of tonnage, within a block of sucrose percentage. Nevertheless, many sugar mills have successfully inducted sugar rich varieties in their respective areas.

12.7 SOIL REQUIREMENT

Sugarcane is well adapted to **wide range of soils** from sandy to heavy clay soils–if water, drainage, fertility and depth are not the constraints. But, the sugarcane yields are always better in medium black soils of Northern Karnataka, Andhra Pradesh and South/mid Maharashtra as well as alluvial soils of Punjab, Haryana and Uttar Pradesh. Sugarcane roots prefer to be well aerated and grow deeper. Hence, the depth of weathered zone is crucial (min: 60 –80 cm). Hard pan, subsoil lime band or salt zone, poor drainage reduce the growth and yields. But surface salinity is tolerable (up to 1.7 dSm^{-1}). The tolerable pH range is 4–9 (with yield reduction), but ideal pH range is 6–8. Sugarcane needs **high fertility** in soil, as it mines large quantity of nutrients during its year long duration. **Shallow alfisols** are not suitable.

12.8 CLIMATIC REQUIREMENT

Sugarcane is more adapted to tropical climate with year round sunshine than subtropical regions, where it was originated. If moisture is not limiting, it prefers the **hot weather** with **bright sunshine** throughout the year. It is a thermo sensitive, photo sensitive **short day plant** (photo synthesis sensitive to temperatures and day length and flowering sensitive to photoperiod). Table 12.1 shows the ideal climate for sugarcane.

- **Temperature:** Growth rates and sugar accumulation are governed by air temperatures. Tillering, root and shoot growth are favoured by 33–36°C, while carbon assimilation, sugar synthesis and its transport are best achieved at 30–33°C. But ripening (conversion of glucose into sucrose) is best achieved at 20–25°C. Temperature below 20°C may decelerate the growth.

- **Sunshine:** For maximum productivity, crop needs bright sunshine hours of more than 10 hours throughout the duration, except last two months (8–10 hours). For this reason, yields are higher in tropics than subtropical regions, where sunshine hours go below 8 hours well before solstices. Premature arrowing, poor recovery (due to lesser sucrose deposition) consequent to poor light are common, when crop is exposed to < 8 hour/ day light.
- **Altitude:** Even though sugarcane can be grown in altitudes between 0–1550 m, the yield is maximum at mid-altitudes of 300–600 m. Higher altitudes reduce the yields (due to reduced sunshine hours) and lower altitudes adversely affect maturity (deposition of sucrose), due to isothermal characters of Indian coasts.
- **Rainfall:** Sugarcane ideally requires around 1500–1800 mm rainfall during growing period of one year. Ideally, this rainfall should be spread throughout the year. In South India, sugarcane cannot be grown as rainfed crop (as South Indian rains are restricted to June–October). In some parts of Uttar Pradesh, it is grown as rainfed crop, even though the crop suffers rainless period for 2–3 months

Table 12.1 Ideal Climate for Sugarcane

For Growth	*For Ripening*
Mean maximum temperature 30–36°C	Mean maximum temperatures 20–25°C
Mean minimum temperature > 20°C	Mean minimum temperature not less than 15°C
Altitude 300–600 m	Altitude 300–1500 m
Rainfall 1500–1800 mm in 10 months	Rainfall nil
Sunshine hours 10–12 hours	Sunshine hours 8–10 hours
Relative humidity 60–70%	Relative humidity 40–50%

12.9 BOTANICAL DESCRIPTION OF PLANT

It is an annual tall growing grass belonging to family *Poaceae*, characterised by the number of side shoots from the basal nodes of main stem (called **tillers**) (Figure 12.1).

- **Stem:** The main stem as well as tillers are popularly called **cane** or even **stalk**. The main stem grows from a bud of a piece of seed stalk (often called **sett**) under ideal moisture. The canes grow vertically to a height of 2–4 m, sometimes even in zigzag fashion and even horizontally. The fleshy and juicy canes have varying colour ranging from black, brown and green, yellow, purple to red. Epidermal layer may be hard or soft.
- **Leaves:** Each stalk can have 1–1.5 m long and 30–100 mm wide 10 –18 lanceolate leaves, with characteristic rough lamina, sometimes serrated and may or may not have trichomes. The leaf sheath surrounds the inter node just below a character of grass family.

- **Inflorescence:** It is called **arrow**. It is an open branched panicle (Figure 12.2) with hundreds of tiny silky flowers, which seldom result into seeds. When fertilised, it is called **fluff**. Seed set is a major constraint in breeding. Many of the varieties do not flower, and if flowering occurs, anthesis fails. When fertilised, they are cross-pollinated, a rule (anemophily).

Figure 12.1 Tillering in sugarcane.

Figure 12.2 Arrows of sugarcane.

Why Sugarcane Breeding Institute (SBI) is Located in Coimbatore?
Many varieties fail to flower and set the seeds. Some of them need artificial inducement to flower. But, in Coimbatore, located at 77°E longitude and 11°N latitude, the temperatures and altitude are ideal for **natural flowering** of most sugarcane varieties. This facilitates **crossing programs** and varietal improvement. Hence, SBI is located in Coimbatore.

- **Roots:** Just below every node on stalk, a **root band** with primordial zone appears. They are responsible for the initial roots called **sett roots**. When sett is placed in moist soil, they germinate and give nourishment to growing shoot from bud and last for 1–2 months, till shoot roots are developed from the base of first inter node. First set of shoot roots called **buttress roots** are thicker, grow rapidly and penetrate deeply and are responsible for anchorage. Second set of shoot roots are fibrous and profusely branched, produced from lower most nodes. They are delicate, thin and responsible for absorption. In addition, 1–2 thicker roots may also be developed (called **rope way**),which grow laterally and provide anchorage. **Sugarcane is considered as deep rooted crop, although, many other grasses have shallow roots** (absorbing zone upto 80–90 cm).

12.10 SUGARCANE GROWING ZONES

Sugarcane Breeding Institute, Coimbatore has divided country into five sugarcane growing zones, based on climate, soils and sugarcane growing areas. They are:

- **North western zone:** Punjab, Rajasthan, Haryana, Uttar Pradesh and Gujarat.
- **North eastern zone:** Assam, Bihar, Nagaland, West Bengal and Odisha.
- **North central zone:** Madhya Pradesh. (All these zones have sub-tropical climates)
- **Coastal zone:** Coastal regions of Andhra Pradesh, Karnataka, Tamil Nadu, Odisha, Maharashtra and Gujarat.
- **Penninsular zone:** Karnataka, Andhra Pradesh, Tamil Nadu and Kerala. (All these have tropical climates)

Temperature, day length, relative humidity, and latitude altitude decide the sugarcane yields, besides latitude with a thumb rule that lower latitudes have higher productivity and vice versa.

Why Productivity is Higher in South India than North India?
• Latitudes are lesser in South India than North Indian states. • Day length for most part of the growing period is more than 10 hours in South India. • Minimum temperature hardly reduce below 20°C during growing period in South India. • Most of the sugarcane area is irrigated in South India, which favours use of more inputs. • Temperature during maturity is in the range of 20–25°C in South India, while it is 8–15°C in North India.

12.11 VARIETIES

Sugarcane is a vegetatively propagated crop. To replace conventional varieties grown prior to 1920, large number of varieties were developed by the Sugarcane Breeding Institute, Coimbatore, which was established in 1912. Hybridisation, selection and introduction at SBI, Coimbatore as well as its sub centres like Karnal, Conoor, Agali (Kerala), Jalandhar, Navasari have led to the evolution of high yielding varieties with desirable traits. They are meaningfully prefixed as Co (bred at Coimbatore), CoN (bred at Coimbatore, selected at Navasari) and CoJ (bred at Coimbatore, selected at Jalandhar). These varieties are recommended and grown in different states, according to the variations in planting season/crushing season, agro climatic features, and soil characters. Important sugarcane varieties released in India are:

- **Maharashtra:** Co 419, Co 775, Co 671, Co 740, CoM 7125, Co 87044, Co 87025, CoM 88121
- **Karnataka:** Co 6415, Co C671, Co 740, Co 87044, Co 8371, Co 8704, Co 86032, Co 62175
- **Haryana:** Co 89003, V = Co J 64, Co 1148, CoS 767, CoH 35, CoH 119, CoH 110
- **Tamil Nadu:** CoC 671, CoC 8001, CoC 9361, Co 6304, Co 85019, Co 86010
- **Uttar Pradesh:** CoS 8436, CoS 95255, CoS 767, UP 22, CoS 94257, CoPt 90223
- **Punjab:** CoJ 64, CoJ 83, CoJ 88, CoS 8634
- **Andhra Pradesh:** Co 6907, CoT 8201, Co7219, CoC 671, Co 62175

Some varieties are specifically bred for specific problematic soils like saline, and water logged soils. Few are breads for high sugar varieties.

12.12 PLANTED CANE AND RATOON CANE

Sugarcane is the only crop, where very successful ratoon is grown for many seasons, after initial planting takes place. Ratoon refers to **allowing the cut shoots to grow into another crop,** after the harvest of the main crop. Commercial cane cultivation involves both **planted cane** (freshly planted by using pieces of stalk called **setts**), which is also called **main crop** as well as **ratoon cane** (after the harvest of the main crop, cut shoots are allowed to be grown), frequently called **ratoon crop**. Ratooning for many seasons (up to 20–25 seasons) is reported, but ratooning for 5–6 seasons is normal. Table 12.2 shows the advantages and disadvantages of rationing sugarcane.

Table 12.2 Advantages and Disadvantages of Ratooning Sugarcane

Advantages	*Disadvantages*
Savings in cost of seed materials (setts), treatment of setts	Reduced cane yield, poor juice quality
Savings in planting cost and preparatory tillage	Continued diseases/pests from the main crop
Least disturbance to biological activity, including earthworms	Difficulties in removal/management of trash after the harvest of the main crop
Matures earlier than planted cane	Ununiform growth of tillers/poor germination

12.13 SUGAR MILL MONITORED PLANTING SEASONS

Sugarcane cannot be disposed of by the farmers in any open market. Dependence on sugar mills for the purchase of canes coupled with limitations of such sugar mills in handling the large tonnage of cane at a time has necessitated the mills to monitor the planting time so that cane supply is in tune with daily crushing capacities. Further, this staggered supply of cane facilitates running of sugar mills for extended period of 240–250 days, without handling dried cane. The cane supply to the mills will also have to consider supplies from ratoon crops. But, maturity of cane not only depends on the age of crop, but also a combination of dry weather and low temperature. When weather changes unfavourably, either the recovery or cane supply is affected. Such a scenario is also the reason for poor recovery in many regions, as the cane will have to be inevitably harvested to meet mill's demands even when it is premature or overmature.

12.14 PLANTING SEASONS

The planting season of sugarcane is different in different parts of the country, either to adjust to soil/climatic limitations or to ensure the supply of cane for longer period.

- Conventionally, annual cane (called **eksali**) of 11–12 months planted in different months like:
  - Spring planting as eksali during (cane matures in 11–12 months)
    - Feburary–March in North India
    - January–February in peninsular India

- Autumn planting during
  - September–October in Northern India (except Bihar)
  - October–November in Bihar and peninsular India

  (such planting is called **pre-seasonal planting** and the cane matures in 13–15 months)
- Late season planting during March–April in North Western India, Madhya Pradesh and Uttar Pradesh after the harvest of wheat. The cane matures in 9–10 months.

- Monsoon planting as **adsali** during July–August in Maharashtra and Northern Karnataka. Such cane matures in 16–18 months.

Why July Planted Cane does not Flower in Ensuing December?
Although, cane maturity depends on dry weather and cold temperatures found in the months of November and December, July or August planted cane does not flower in forthcoming months of November–December, because adequate vegetative growth would not have taken place by such time. The crop has to wait for next November–December to mature and convert its glucose reserves into sucrose. Hence, such crops take more time to mature and yield higher cane yield and even better recovery percentage. Adsali and pre-seasonal canes occupy the land for 13–18 months and have advantage of more photosynthesis and accumulate more sugars, grow thicker canes and bear more tillers. Hence, their yields are always higher than annual (eksali) or late season plantings.

12.15 PLANTING METHODS

- **Flat bed planting:** It is followed in Uttar Pradesh, Haryana, Punjab and Rajasthan states. Shallow furrows are opened in soils of low fertility at a distance of 60–90 cm by wooden plough. Setts are planted end-to-end and covered with soil (5–10 cm). Setts can overcome the effect of cold temperatures in autumn planting.
- **Trench planting:** It is followed in Uttar Pradesh, Bihar, West Bengal, Andhra Pradesh and Odisha. Trenches of 25–30 cm deep and 30–40 cm wide are opened in highly fertile soils at a distance of 90 cm. Setts are placed end to end and covered with shallow layer of soil (5 cm).These trenches are irrigated initially. Trenches are arranged in line to facilitate later management, including irrigation later as furrow irrigation.
- **Ridges and furrow planting:** It is followed in Maharashtra, Karnataka, Andhra Pradesh and Tamil Nadu. Furrows of 30–40 cm deep are opened at 90 cm distance. Setts are placed end-to-end on the top of ridge and slightly pressed on them. Furrows are irrigated.
- **Paired row planting:** This is a recent system to facilitate mechanised harvesting followed in peninsular India, also called **wide row planting**. Keeping the distance between ridges up to 5 feet, two rows of cane are established (which are 1 foot apart) on wide topped ridge. The two rows of canes are placed end-to-end on ridge and irrigated by separate furrow for each crop row. This system more suits drip irrigation, wherein the lateral can be laid in between two rows.
- **Partha planting:** It is principally followed in coastal heavy rainfall regions of Andhra Pradesh and Tamil Nadu. Three budded setts are planted on ridges at 60° angle in such a way that one older bud is inserted in soil and other two younger buds are open in air.

After rain recedes, they are irrigated. By such time, young shoot would gave come out from younger bud and older bud would have established roots.

- **Skip row planting:** It is followed in Odisha. It is the modified trench method, wherein continuous trenches of 45 cm width and 15–20 cm deep are made in paired order in such a way that each pair of trench rows are placed 90 cm apart. Setts are placed end-to-end in each trench and soil is covered on them. After emergence, the planting results into paired rows of 45 cm, each pair located 90 cm apart.
- **I.I.S.R. method:** Trenches of 90 cm wide are opened at a distance of 20 cm between two trenches. They are filled with FYM and litter and 3–4 budded canes (called long rayunguns) are planted vertically with intra row spacing of 30 cm.
- **STP method:** It is a technique developed by Indian Institute of Sugar Research (popularly called STP method–**Spaced Transplanting Technique**). A nursery bed (50 m^2) area is sufficient for 1 ha) is prepared and single budded setts are planted closely and vertically in such a way that bud is outside. The seedlings are ready after 5 weeks. These are transplanted in the main field at a distance of 45 cm in each row which are 90 cm apart. This method helps to improve the germination percentage (85–95%) as well as reduce seed requirement to 2 t/ha as compared to 6–7 t/ha, required by regular method.

Advantages of STP method
(1) Saving of 4 tons of seed cane/ha (2) Uniform crop stand with higher yield (3) Reduction in late shoot production (4) Reduced cane lodging (5) Increase in seed multiplication ratio (1:40 as compared to 1:10) (6) Higher stalk population/ha

- **Ring system:** To encourage mother shoots, this system involves digging of pits of 90 cm diameter in 120 cm apart rows and filled with FYM. Each pit is planted with 3–4 budded 20 setts. The germination is as high as 80–90%.
- **Machine planting:** In recent days, the cane is planted by using specialised machines developed to drop the setts at required spacing, as the machine moves in row. It saves the labour and can cover large area in short time. They are gaining popularity in many sugarcane growing areas.

12.16 PLANTING MATERIALS

- **Setts:** Most commonly used seed material for planting is 1/2/3 budded immature piece of cane called **sett** (Figure 12.3). They are obtained by raising a separate seed cane (6–7 months–entire cane could be used as sett) or by cuttings from top one-third of mature harvested cane.
- ***Preparation for setts*:** It is desirable to grow separate sugarcane crop for seed materials, rather than using one-third top of harvested cane for better germination and cane yield. Such crop is harvested at 6–7 months, when entire plant can be used for sett preparation and entire cane is filled with glucose. In later stages, this glucose is converted into sucrose, which is not desirable for germination as growth. **Due to apical dominance, it is not desirable to use entire cane or more than three budded cane**

as planting material. If such material is used, the germination of lower buds will take an extended time and the stand of crop will be ununiform. Seblang or tiller separation can also be adopted to produce setts.

Figure 12.3 Single budded sugarcane setts.

It is desirable to treat setts with moist hot air treatment at 54°C and 99% humidity for 2 to 2.5 hours to minimise diseases like red rot, smut, ratoon stunting disease, grassy shoot disease and leaf scald. In addition, it is useful to treat the setts with 0.5% Agallol solution for half an hour.

- **Rayungans:** The top of the mature cane is cut off to facilitate auxiliary buds to sprout (due to apical dominance). Then cane pieces with 3–4 buds are planted in portion of field selected for rayungan production. After sprouting, they are cut into single budded rayungans. They are transplanted after 3–4 weeks. (**Rayungan** is Indonesian term: this technology is adopted in South Eastern Asia).
- **Seblang:** Setts are planted in fertile soil at wider spacing. Immediately after they germinate and tiller (after 2–3 months), the tillers are separated along with roots and used for transplantation. (practiced in Cuba and Java)
- **Pre-germinated setts:** In Spaced Transplantation Technique of planting, setts germinated in nursery are used. They are also called rangoons in Kolhapur.

> Apical dominance is a phenomenon indicating dominance of apical buds than other lower buds due to the presence of auxins. When apical bud is cut, next bud is activated.

- **Tissue culture:** Meristem tips are excited and grown in an independent artificial medium under controlled condition. This is not used for commercial cultivation, but adopted to revive degenerated varieties.
- **Tjeblocks:** The whole cane is cut in middle at about 6–7 months and upper part is planted vertically with the lowest node in soil for rooting. Such planted pieces of cane are called **tjeblocks**. After 20–25 days, single budded setts are prepared from both tjeblocks and mother cane and used as rayungans. This is improved rayungan method.

- **Buds and chips:** Buds on each stalk are scooped along with little flesh and planted in small polythene sleeves with manure and soil and allowed to sprout for 10–15 days. The sleeves will have holes below for rooting. They are used for transplantation in main field.

12.17 CROPPING SYSTEMS

Sugarcane is a long duration crop occupying the land for 10–18 months. Hence, two-year and three-year rotations are practiced, wherein sugarcane is rotated with green gram, wheat, potato, groundnut, finger millet as well as rice. However, most popular sequential system is sugarcane–ratoon.

More popular system is companion cropping with potato, rajmash, green gram, soybean, French bean, groundnut, onion, linseed, and amaranthus, depending on the suitability of the crop. These crops are short duration crops, which can be grown along with initially slow growing sugarcane as companion crops. Most of the cost of cultivation of sugarcane can be compensated by such crops, so that sugarcane cultivation is more profitable. Sugarcane does not offer competition to these crops, as it is slow growing; nor is its growth affected by these crops.

12.18 NUTRIENT MANAGEMENT

Sugarcane is a heavy feeder and extracts more nutrients than most of annual crops. It takes up around 140–160 kg N, 40–160 kg P_2O_5 and 200–220 kg K_2O per hectare at around 180–200 days duration, after which it will decline, depending on the number of soil, crop and climate-related factors. Nutrient sufficiency concentrations for nitrogen are 24–25 g/kg, and 19 g/kg respectively at 3 and 6 month old crop, while it is 2–3 g/kg for phosphorus and 12–20 g/kg for potassium at 3–6 months age.

Fertilisers could be computed based on number of testing methods, each of which has unique criteria. Few such methods are:

- **Soil test crop response:** Considers yield target oriented nutrient requirement duly considering the losses in each of the soil type.
- **Crop logging:** It is a dynamic approach, designed to maximise the cane yields, and made up of pertinent data on chemical and physical observations of the crop which reflect the health of the crop. It involves periodical analysis of plant samples and physical examination of the crop and decides the nutrient needs.
- **Foliar analysis:** Changes in composition of plant tissues are taken as quantitative index of nutrient demand. The fertilisers are decided on the basis of foliar analysis

Although, fertiliser needs are judged based on the variety, soil fertility status and rainfall/irrigation availability in each situation, it can be generalised that sugarcane needs to be applied with 112–224 kg N/ha in North Indian situations and 112–400 kg/ha in tropical situations. Considering that sugarcane can utilise 20–25% of applied phosphorus, optimum dose of phosphorus to be applied varies from 50–80 kg P_2O_5/ha in North, 30–90 kg P_2O_5/ha in South, 120–180 kg P_2O_5/ha in West and 60–100 kg P_2O_5/ha in Eastern India. Even potassium levels vary from 30–185 kg K_2O/ha, although they also vary in different regions.

In general, phosphorus and potash fertilisers are applied as basal, although, in some states like Andhra Pradesh, even potassium fertilisers are applied in two splits (half before planting + half after 90 days). But nitrogen fertilisers are invariably given in split doses. In subtropical India, half of recommended N is supplied at the time of planting and remaining after 6 months. Whenever the crop received higher N doses (more than 180 kg N/ha) second half dose is further split (one-fourth after 3 months + one-fourth after 6 months). But in Bihar, one-third dose is given as basal and remaining one-third dose after 45 and 90 days after planting is given. In Karnataka, basal N dose is as small as one-tenth dose followed by remaining dose every month till 6 months.

Sugarcane is benefitted by the application of *Azotobacter* (@5 kg/ha along with 4 tons press mud/ha), in terms of saving 25–35 kg N/ha. Similarly, use of *Acetobacter* has not only reduced N requirement, but promoted the growth due to release of growth promoting substances. Use of Phosphate solubilising bacteria like *Bacillus* or fungus like *Aspergillus* has also helped in reducing the phosphorus application.

Application of sulphur @50–100 kg/ha in sulphur deficient soil, zinc sulphate @25 kg/ha may be necessary in some soils. Similarly, lime induced chlorosis can be corrected by the application of 25–50 kg ferrous sulphate/ha.

12.19 WATER MANAGEMENT

Ridges and furrow method of irrigation is most common practice to irrigate sugarcane. When water is in short supply, **skip row method** of irrigation is more ideal. In paired row normal planting/wide row planting, **drip method of irrigation** is popular when source of irrigation is well. Otherwise, furrow irrigation is followed. In recent years, subsurface drip irrigation is adopted by the farmers, wherein the lateral in drip tubes are underground (10 cm deep) and are permanently left for next 4–5 ratoons. The water savings by subsurface drip irrigation is significant even over regular drip irrigation

In general, irrigation at 0.7 IW/CPE ratio is ideal. During monsoon, the tropical regions need 6–12 irrigations, while irrigations may not be necessary in subtropical regions–unless long rainless period is witnessed. But, in post-monsoon period, 6–8 irrigations are necessary in subtropical region and number of irrigations in tropics may even go up to 5–15 irrigations. Last 4 weeks should not be irrigated to facilitate maturity. **Tillering**, **cane elongation** and **early ripening** are critical stages for soil moisture. Water requirement of sugarcane is as high as 190–220 cm in tropical regions, while it is 150–200 cm in subtropical region.

12.20 WEED MANAGEMENT

The weeds can reduce the growth of cane significantly and reduce the yields, as the crop growth is slow in the initial stages. The crop weed competition is critical even up to 120 days and even up to 150 days in some situations. The options to control weeds include **frequent intercultivations** both ways—as the spacing is 90 × 60 cm. In wide spaced planting, it is much easier to take up intercultivation. Second option is to **grow companion crops** for first 60–90 days, so that crop experiences less weed competition. The third option includes mulching the sugarcane trash from the planted cane after the harvest for ratoon crop which serve the purpose of moisture conservation as well as weed control. Of course, pre-emergence spray of Metribuzin (1–2 kg/ha) or Alachlor (1–1.5 kg/ha) or Atrazine (1–2 kg/ha) can be applied to reduce the weed population significantly. Even post-emergence spray Prosulfuron (20–30 g/ha) or Metasulfuron methyl are useful to reduce the weed population up to 120–140 days.

12.21 OTHER CULTIVATION PRACTICES

- **Land preparation:** Deep cultivation is necessary for all planted canes, achieved by using MB ploughs and disc ploughs. It must be followed by clod crushing, harrowing and levelling. Application of 10–15 t/ha organic manures is necessary. Then the land is laid out into either shallow ridges, or deep ridges or broad ridges or left as flat bed–depending on the method of planting.
- **Seed rate:** Depending on the planting method and spacing adopted, the number of setts required to plant one hectare of land varies from 25,000–30,000 (Andhra Pradesh, West Bengal, Maharashtra, Karnataka, Gujarat, Kerala, and Madhya Pradesh) to 38,000 –40,000 (Assam, Bihar, Odisha and Uttar Pradesh) to 40,000–45,000 (Rajasthan), 50,000 (Punjab, and Haryana) and even up to 75,000 (Tamil Nadu)—the number get increased even up to 88,000–1,00,000 setts/ha. Hence, the weight of setts varies from 6 tons to even 15 t/ha.
- **Spacing and plant population:** In most of the situations, 90 cm spacing between rows and 60 cm between plants is desirable with a population density of 54,000/ha (each sett has three buds resulting into three main shoots). In wide spaced double row planting, spacing between rows may be 30 cm or 45 cm and between pairs of rows, it can be 5 feet.
- **Propping:** It is a special practice adopted in sugarcane–involving tying the 5–6 tillers together with bottom leaves at a height of 2–3 feet above the soil, so that they do not fall/lodge and reduce the cane yield. It is followed after 9–10 months of planting.
- **Harvesting:** Harvesting is usually done by cutting the canes at bottom, detrash them (removal of dried leaves), and removal of top one-third part of cane. The top portion has less sucrose and more glucose (for this reason, it is used as planting material). So, prepared **millable canes** (Figure 12.4) are bundled in a group of 12–15 canes. Only

Figure 12.4 Canes ready for crushing (millable canes).

such canes are sent for milling. Generally, the cane yield is expressed as millable cane yield. Then they are transported to sugar mill or crushed at the site of cultivation for preparation of jaggery. Recently, mechanised harvesting by use of specialised machines has become very popular. Such machines cut the canes, detrash them and even reject the cane tops. Depending on the varieties, availability of irrigation and management conditions, cane yield of around 70–130 t/ha can be expected.

12.22 RATOON MANAGEMENT

At any given point of time, nearly 55% of sugarcane area is occupied by the ratoon cane, although, it contributes 30–35% to production, indicating that its yield potentiality is lower than planted cane by 20–25%. Nevertheless, rationing is most common in commercial cultivation to save the cost of planting, planting material as well as to save time (ratoon crops mature earlier). The management of ratoon crop requires many specific practices. They are summarised as follows:

- **Time of ratooning:** Obviously, emergence of ratoon crop depends on when planted crop is harvested, which in turn depends on the permit issued by the sugar mills. Ideally, ratoon crop germinates well at 25–30°C. In tropical states, December to March harvest of planted cane has resulted in good ratoon crop, while in subtropical belt, spring (February–March) cut of planted crop gives better crop of ratoon.
- **Trash removal:** Harvest of previous crop leaves large quantity of dry trash, almost covering the entire land. It is estimated that each hectare generates trash of 10–12 tons of trash consisting of dry leaves and rejected cane tops. They have to be removed manually to cultivate the land and grow ratoon. Most farmers burn the trash in the field to avoid spending on labour to collect trash. However, this will reduce the percentage of germination and burns the valuable organic carbon in soil. It is more desirable to clear the cut portion of cane and retain the trash in between rows to act as mulch to conserve water and reduce the weed population. The trash can be broken to small pieces by trash pulveriser and may be allowed to decompose *in situ*, assisted by cultures like *Pleurotus* or *Aspergillus*.

Disadvantages of Trash Burning	Advantages of Trash Burning
1. Loss of organic carbon by high temperatures	1. Kills insect pests/pathogens
2. Reduction in bud sprouts in ratoons	2. Cultivation is easier
3. Expose the land for evaporation	3. Adds the ash to the soil

- **Off–barring and ridge flattening:** If the soil has compacted by harvest time, ratoon crop will be severely affected. To improve the physical conditions, off barring and ridge flattening is done. This is also called **shoulder breaking**. They are achieved by cutting the ridges on either sides. This helps in quick emergence of newer roots from stubble and helps in decay of old roots and even trash. This operation is done by tractor drawn shaver or even ridger after 10–12 days of harvest.

- **Stubble shaving:** Cutting the canes at ground level during harvest of main crop is very important for the success of ratoon. Stubble shaving after the harvest involves reducing the height of cut if harvest is not done at the ground level. Stubble shaving also involves removal of excessive number of side shoots and retaining 8–10 shoots per clump. Both these operations are done by sharp sickles or tractor drawn stubble shaver. The objectives of stubble shaving are to expose and facilitate healthy underground buds to germinate and establish the deep root system than above ground buds.
- **Gaps and plant population:** Gaps in ratoon crop varies from 10–30%. However, ideally gaps should not be more than 15% for a good ratoon crop. Gaps could be filled by using seedlings raised in poly bags or nursery. Trash mulching stimulates the bud growth and reduces the gaps, while burning the trash seriously reduces the plant population in ratoon crop.

Advantages of Trash Mulching
1. Reduces the weed population
2. Helps in better bud germination
3. Conserves the moisture
4. Postpones irrigation
5. Improves the biological activity
6. Helps in better decomposition

- **Fertiliser application:** Ratoon crop should receive 25% more nitrogen than planted cane, as the crop will have shallow root system, nutrients would have immobilised and extra need from sprouting of buds. In general, around 200–300 kg N/ha with 60–80 kg P_2O_5/ha and 60–100 kg K_2O/ha are recommended for ratoon crop. In North India, entire dose of fertilisers is applied immediately after stubble shaving, while in South India, nitrogen is supplied in two splits (after stubble shaving and after 45 days of stubble shaving).
- **Irrigation:** The ratoon crop should be irrigated once immediately before off barring and subsequently after stubble shaving for better germination of buds. Subsequently, it is irrigated like main crop.
- **Earthing up:** After 3–4 months of ratoon crop, the earthing up operation has to be carried out.

12.23 PESTS AND DISEASES

- **Red rot:** Most destructive disease, caused by fungus *Colletotrichum*, carried from year to year through infected setts, and debris. Its spores remain in field after harvest. Symptoms include loss of colour, withering of 3rd or 4th leaves from top, drying, wrinkling of stalk, which becomes hollow. The pith of stem is seen red with transverse white bands on the opening of stem. Avoiding ratooning, use of disease free setts, dipping setts in 0.25% solution of Agallol are some measures to control or avoid the disease.
- **Smut:** The disease is caused by *Ustilago*, whose damage is more severe from April to July, especially in ratoon crop. It can reduce juice yield by 10–15%. Symptoms

include black whip like structure at the apex, thinner stalk, and buds turning papery. The diseased clump is more tall and has more tillers than normal clumps. Use of disease free setts, removal of smutted whips in covered bag, avoiding ratoons are few measures to control the disease.

- **Wilt:** Disease symptoms include yellowing and withering of crown leaves, followed by rapid drying of cane associated with pith of stalk turning reddish brown with no white streaks. Even rind can shrivel. Control measures include selection of good setts, removal of affected canes and avoiding rations as well as dipping setts in 0.25% Agallol solution are few control measures suggested.
- **Grassy shoot (Albino):** It is caused by virus and the disease symptoms include production of numerous thin tillers bearing pale narrow leaves resembling grasses, with a characteristic reduction of intermodal length. Uprooting the affected plants, hot air treatment of setts, at 54°C for 8 hours and avoiding ratoons are some measures suggested to avoid/control the disease.
- **Ratoon stunting disease:** The disease caused by virus, is characterised by the stunted growth with thin short canes and pale leaves. On maturity, such canes may develop pink colour inside the stem. The disease is spread through setts and sett treatment with hot air at 54°C and 99% humidity for 2–3 hours is effective to control the disease. Avoiding the ratoon crop is necessary to stop spreading it.
- **Shoot borer:** The insect attacks the crop in the earlier stages (2–3 months crop). The caterpillar eats away top part of whorls and soft parts of apex and results in 'dead hearts'. Unopened central shoot rots and when pulled, gives away rotten smell. Such plants totally die or when plant does not die, the more number of weak tillers are initiated. Frequent irrigation creating high humidity, spray of Monocrotophos @2.5 ml/l can reduce the damage to some extent.
- **Root borer:** The caterpillars bore at the base of stem and make tunnel downward and feed on roots, causing the dead hearts, which cannot be pulled out easily. It can be controlled by measures suggested against shoot borer.
- **Top borer:** Larvae enter the midribs of top leaves and make tunnels downwards toward the central growing point. Tunnelling of top part of shoot creates dead heart, but associated with holes on the leaves and red galleries in the midrib of top leaves. Application of 30 kg Carbofuran or the spray of Monocrotophos @1.5 ml/l will reduce the damage.
- **Pyrilla:** Both nymphs and adults suck the sap from leaves and attached spots on the leaves turn to yellow. The attack reduces the vitality of leaves and reduces the photosynthesis and thereby reduces the biomass production. They are more effectively controlled by releasing their parasites *Epiricania melanoleuca* @1500 beetles/ha or 40,000–50,000 eggs. In case of failure, Malathion could be sprayed @1.25 ml/l.

12.24 RIPENING AND MATURITY

Sugarcane matures after the ripening phase, which is the process of sucrose accumulation preceded by the conversion of glucose into sucrose. Ripening is influenced by climatic features, age of the cane as well as many management factors like inter cropping, lodging and fertilisers. Harvesting mature cane gives maximum recovery percentage as well as maximum sugar/jaggery yield. Identification of maturity is crucial for the success in sugar production, as under mature canes and over mature canes pose similar problems of low recovery.

Onset of cold season, or reduction in temperature of 20–25°C is invariably favourable to maturity, if associated with dry weather. A sugarcane crop of more than 6–8 months will mature, when it is exposed to such climate–irrespective of planting time. But cane of less than 6 months may not mature, even when it is exposed to such climate. The ripening process is also assisted by no irrigation in the last 30–45 days of the crop. When, the sucrose is converted back into glucose due to over maturity, it is called **inversion**. The sugar/jaggery yields are the function of sucrose in the cane and hence, over matured cane should not be harvested. Even when harvested sugarcane is left uncrushed for more than 2–3 days, inversion may set in and reduce the sugar yield. Normally, a mature cane should be able to give a sugar recovery of 10–11% under ideal climatic and crushing condition and around jaggery recovery of 9.5–11.5%.

Sugarcane farmers believe that cane matures after flowering. The process of flowering (arrowing) is also governed by the age and climatic features like maturity, hence they may be correct partially. Sometimes, the process of arrowing itself may get extended, leading to harvesting over mature canes. Generally, if the arrowing takes 5–8 days, it is correct to harvest the canes within 10–15 days of arrowing. The delay in harvesting after 15 days of arrowing may lead to reduced sugars. Sometimes flowering is misleading, as some varieties flower well before the onset of cold season, when maturity has not yet set in. In fact, arrowing itself is considered as deterrent to increasing sugars in the cane, as un-flowered cane has higher sugar yield. Attempts are made to suppress the flowering by chemicals (called artificial ripening).

Artificial ripening is also attempted by the use of chemicals like Ethrel at 1–1.5 kg/ha at the beginning of milling season or Polaris @1–1.5 kg/ha 4–8 weeks, before the harvest or even Cycocel @4 kg/ha 50–70 days before the harvest. Even well-known herbicide Glyphosate is popularly used as artificial ripener. Artificial ripeners are also called **sucrose boosters**—as they improve sucrose percentage.

12.25 QUALITY CONSIDERATIONS

The maturity is best indicated by testing the cane by **Brix refractometer**. It is hand held portable instrument to measure total dissolved solids. The total dissolved solids increase with age and is maximum at maturity. The ideal Brix readings to indicate harvesting time range 19 –24. A drop of juice from freshly cut cane is placed on refractometer plate and read against sunlight to get instant readings. But correction factor needs to be applied, as the readings are calibrated for 20°C.

As the crop matures, the moisture content reduces, sucrose content increases, reducing sugars and non-sugar solids reduce.

Important quality considerations are:

- **Brix values:** A good mature cane have Brix values between 19–24 and such cane should give high sugar recovery and high juice recovery.
- **Pol in juice: Polarimeter** readings of sample juice (pol in juice) give more precise reading of sucrose contained and is a better measure of quality. These values vary

between 11–16%. However, polarimeter readings in cane (pol in cane) can be lesser than pol in juice as total weight of cane is considered.

- **Purity percentage:** It indicates the share of sucrose in total dissolved solids. It is obtained by dividing pol in juice with corrected Brix % of juice, expressed as percentage Purity % = (pol in juice/corrected Brix values) × 100.
- **Recovery percentage:** It indicates quantity of jaggery or sugar obtained per unit of cane weight. It is calculated by:
  - ❖ Recovery % = (weight of sugar or jaggery obtained/weight of cane crushed) × 100

  It may also be obtained by:
  - ❖ Sugar recovery % = [S–0.5 (B–S)] × 0.73, where, S = sucrose % (pol in juice) and B = corrected Brix values
- **Commercial cane sugar (CCS):** It is a measure of sugar obtained per unit area from the millable cane harvested. It is obtained by: CCS (t/ha) = Recovery % × weight of millable cane (t/ha)
- **Crystal sugar:** Bright white to dull white colour; crystalline freely falling; crystal size 0.5–3 mm; 2–3 years keeping quality; sucrose content 98–99%; 100% solubility.
- **Jaggery:** Dark brown to light crimson (natural) or light yellow to light brown or whitish yellow to dark golden coloured; various shaped–bucket shape, cubical shape, powdered, flakes, hand shaped and even liquid jaggery is available; sucrose–70–80%, balance made of moisture, minerals and vitamins; keeping quality 2 months to 1 year; hardness depends on the setting point or mineral content of juice.

12.26 TERMS USED IN SUGARCANE

- **Sett:** Piece of sugarcane stalk having 2–3 buds on it, used for planting the fresh crop.
- **Arrowing:** The process of initiation and growth of inflorescence in sugarcane.
- **Fluff:** Fertilised mass of seed of sugarcane, often in mass of fibrous awns.
- **Sett roots:** First set of roots evolved from the **root eyes** at the base of each sett, when ideal conditions are provided at the time of planting. They are temporary in nature.
- **Bagasse:** Fibrous part of cane left over after the juice is extracted.
- **Detrashing:** Process of removal of dried leaves at the time of harvesting.
- **Water shoots:** Late grown shoots, which do not accumulate sucrose at the time of harvest and pose problems of recovery percentage. They are expected to be removed or not harvested.
- **Millable cane:** That part of cane which is obtained by cutting top one-third part of the harvested cane (to remove glucose rich part) and detrash it to make it ready to be crushed.
- **Planted cane:** Sugarcane obtained by planting the setts freshly.
- **Seed cane:** The cane raised specifically for the purpose of obtaining setts for propagation.
- **Stale cane:** The cane not used for crushing, as the cane would have deteriorated in terms of its sucrose content.

- **Inversion:** Process of conversion of sucrose back into glucose, when the overmature, cane is harvested or harvested cane is kept uncrushed for more than 1–2 days.
- **Eksali:** Sugarcane crop planted during normal season maturing in 10–12 months.
- **Jaggery:** Solidified sweet blocks obtained by boiling the sugarcane juice.
- **Adsali:** A sugarcane crop, planted 3–6 months in advance and harvested after 15–18 months.
- **Ratoon:** A crop rained from subterranean buds from the stubbles after the harvest of main crop.
- **LTM/TVD leaf:** Last transverse mark/transverse visible dewlap leaf of sugarcane, used as an indicator leaf for foliar analysis of nutrients to know the needs of the crop.
- **Seedling/settling:** Seedling is young plant develop from the germination of true seed, normally used by breeders; settling refers to **young plant coming out of sett**, commonly known to produce commercial cane.
- **Brix:** The values of Brix refractometer, indicating the total dissolved solids, including sucrose. The values of Brix should be around 10–24 at the time of harvest.
- **Pol. in juice:** It refers to **polarimeter readings**, indicating the sucrose percentage in juice. The pol in juice values vary between 11–16% in the juice from the cane ready for harvest.
- **Commercial cane sugar (CCS):** It is sugar manufactured from one hectare of land, expressed as t/ha. It is calculated by multiplying recovery % with cane yield (millable cane).

12.27 JAGGERY MAKING

Jaggery is a solidified product obtained by boiling sugarcane juice. Nearly 45% of Indian cane is crushed for jaggery. Manufacturing jaggery has remained as unorganised cottage industry, primarily taken up by the farmer himself or farmers turned into entrepreneur.

Process of making the jaggery

The millable canes obtained from the field are crushed in roller, by using diesel/electric motor and the juice is collected in a tank after filtration. The juice is then transferred to the boiling pan (made of cast iron) of 8–12 feet diameter, 2–2.5 feet height either through pipes or manually. The juice is boiled for 3–4 hours on country made oven in a thatched shed, invariably using dried bagasse (obtained while crushing the canes for juice). The juice is then clarified by adding lime, super phosphate as well as extract from Okra plant. Frequent removal of impurities as **scum** is necessary for at least 6–8 times in a span of 3–5 hours. Then the juice starts solidification and reaches **setting point** which is around 105–110°C. At such point, the whole contents of pan are emptied to molds of various shape (depending on what shape the final jaggery should have) and allowed to set. The setting time varies from 4–8 hours. Then the jaggery is removed from the mold and allowed to cool and later packed. The whole process from juice extraction to final product may take around 20–24 hours, and the process cannot be stopped in between. The farmers manufacture jaggery in batches, on continuous basis by handling three to four continuously rolling batches.

The technology to manufacture is not standardised by scientific study, but by farmers knowledge. Hence, the jaggery technology varies from region to region and is not uniform. Further, the whole technology appears inefficient, in terms of juice extraction (due to poor crushing ability), and fuel consumption. The experience of person in charge will decide the quality of jaggery. The whole process of manufacturing of jaggery is labour intensive and has become costly exercise, in view of shortage of skilled labours. The recovery percentage varies from 9–11% but it indicates the weight of jaggery and not that of sucrose. The technology need to be refined, standardised and popularised. Moreover, many farmers prepare jaggery in unhygienic condition which leads to the poor quality jaggery reaching the market.

Jaggery and health

Crystal sugar manufactured from sugar mill has 98% sucrose and negligible quantity of other nutrients. But, jaggery has around 75–78% sucrose, 5–6% moisture and minerals like iron, phosphorus, calcium, manganese as well as vitamins. For this reason, use of jaggery is always advantageous to human health than crystal sugar.

REFERENCES

Anonymous (2014), Sugarcane, Wikipedia http://en.wikipedia.org/wiki/Sugarcane.

Anonymous (2014), *FAOSTAT*, http://faostat.fao.org/site/567/DesktopDefault.aspx?PageID=567#ancor.

Anonymous, 2022, Agricultural statistics at a glance, DAFW, GoI, p. 29.

FAOSTAT, 2023, https://www.fao.org/faostat/en/#data/QCL.

Reddy, S.R. (2009), *Agronomy of Field Crops*, 3rd Revised Edition, Kalyani, Ludhiana, pp. 499–557.

Singh, Chhidda, Prem Singh and Rajbir Singh (2003), *Modern Techniques of Raising Field Crops*, 2nd ed., Oxford and IBH, New Delhi, pp. 465–489.

Chapter 13

Sugar Beet (*Beta vulgaris* L.)

13.1 ECONOMIC IMPORTANCE

Sugar beet is an important sugar yielding crop in the world, like sugarcane. Out of 130 million tons of global production of sugar, 35% is generated from the **sugar beet** and 65% comes from the sugarcane. But, in India, contribution of sugar beet towards sugar production is negligible. Sugar beet is not same as popular vegetable crop beet root which is a variation of sugar beet. Sugar beet contains high deposition of sucrose (12–21%) in its roots. Commercial sugar production from sugar beet is a profitable proposition in temperate countries due to its climatic adaptation. Besides production of sugar, a large quantity of molasses is produced during sugar production. The fibrous insoluble content of the sugar beet, rich in cellulose, hemi-cellulose and lignin, is a popular cattle feed.

Sugar beet is used to produce beverages like vodka, rectified spirit, rum and flavoured liquor. It is also used to prepare sugar beet syrup (looks similar to honey), an alternate natural sweetener. Other uses include production of alternate biofuel called **biobutanol** (chiefly in UK). It was traditionally used as cattle feed, before its commercial value was popularised. The variant of sugar beet—beet root is a popular vegetable as well as food colouring agent (due to its rich deep crimson colour by pigment called **betanin**).

13.2 ORIGIN AND HISTORY

Sugar beet is said to have originated in the **Mediterranean coasts of Italy,** where wild types of sugar beet were found. Many countries in the Europe were found to grow sugar beet in many wild forms. Before the present form, sugar beet underwent several genetic modifications in terms of its sugar content, colour, shape of root as well as adaptability. These modifications were assisted by repeated natural hybridisation between native varieties to finally evolve beets with sugar rich, white, conical shaped roots. The **roots of sugar beets** are generally **white** in colour, while **beet roots** are **deep crimson** in colour. Sugar beet, beet root and chard share common ancestor called sea beet (Beta vulgaris maritime).

The name **beta** ascribed to sugar beet is believed to have been originated from the similarity of its swollen roots to the shape of Greek letter **B**. During mid-1700, Andreas

Marggraf discovered that sugar could be manufactured from sugar beet. After about half a century, during post Napoleons period–1810, many factories were established throughout the Europe to manufacture sugar from sugar beet. The importance of sugar beet spread, only after the commercial possibility of producing sugar was established. Many temperate countries including Russia, China started growing sugar beet only in the nineteenth century.

Sugar beet was introduced to India during 1950. Since that time, Indian Institute of Sugar Research, Kanpur has taken number of initiatives to make the crop popular under Indian condition. Under such initiatives, even a beet sugar factory was established at Sriganganagar, besides arranging production of seeds (at an altitude of 5000) developing agro-techniques and identification of areas to cultivate sugar beet and establishment of All India Coordinated Project on Sugar beet in 1971. Despite all these efforts, area under sugar beet remained unrecognisable at national level–although, some area is developed in Punjab, Haryana, Uttarakhand and Himachal Pradesh. The main constraint to expand area under sugar beet is absence of large number of sugar industries to process sugar beet.

13.3 AREA AND DISTRIBUTION

Sugar beet is produced globally over 4.8 million hectare with a production of 273 million tons at mean productivity of 57 t/ha. France (0.38 million hectare), Germany (0.39 million hectare), Ukraine (0.25 million hectare), Russian federation (1.03 million hectare), Poland (0.26 million hectare) and UK (0.1 million hectare) are major producers of sugar beet in the world with Europe contributing to 2.8 million hectare. It is essentially temperate crop of Europe and USA, with USA producing around 11% of the global production. Contribution to global sugar beet by Asian countries is negligible, although, temperate part of China does produce sugar beet. Prominent Asian countries producing sugar beet are Turkey (0.36 million hectare), China (0.16 million hectare) and Iran (0.1 million hectare). India has negligible sugar beet area (FAOSTAT, 2023).

13.4 INDUSTRY INTERFACE OF SUGAR BEET

Sugar beet is a **commercial crop**, and prospects of its cultivation is entirely associated with commercial possibility of using sugar beet on bulk scale–obviously to produce sugar or brew liquor or to produce biofuel. The area under sugar beet can be expanded only, when suitable processing centres are established to manufacture these products from beet. Under the absence of clear policy on providing incentives to establish beet sugar industries, the manufacturing capacities to produce sugar or other products from beet are even now limited in India. In recent years, sugar beet has found an important utility in **manufacturing ethanol**—a major biofuel and made sugar beet as an energy crop. Even this application can boost beet production only when policy supports are provided.

However, technically Indian Institute of Sugar Research, Lucknow (IISR) has proved in the last two decades that sugar beet can be grown in many subtropical regions of India successfully. It has developed many successful varieties, maintained germ plasm, produced breeders' seeds, and development of hybrids, agro-techniques for beet production, handling and processing sugar beet. The technique to grow tropical beet is further refined by TNAU, Coimbatore.

Tropical Sugar Beet
Tropical sugar beet refers to the cultivation of specific varieties/hybrids of sugar beet developed and released for tropical regions. Such varieties/hybrids are successfully grown in September–November and harvested during March–May months. They can withstand higher temperatures (20–25°C for germination; 30–35°C for development and 25–30°C for sugar accumulation. Presently, Tamil Nadu Agricultural University and Vasantdada Sugar Institute in Maharashtra spear head the mission to popularise cultivation of tropical sugar beet.

13.5 SOIL AND CLIMATIC REQUIREMENTS

The crop is generally grown under **rainfed conditions**, but it can also be grown under irrigation in the subtropics. The crop is grown in **temperate climates** and recently in subtropical regions–where its yields are likely to be lesser. Seeds of temperate varieties can germinate even at 5°C, but the ideal minimum temperatures are considered to be 7 to 10°C. Higher temperatures up to 25°C during vegetative growth are preferred for these varieties, but high sugar yields are assured when night temperatures ranges between 15 and 20°C and day temperatures between 20 and 25°C during late stages of growth. The temperatures greater than 30°C greatly in later stages decrease sugar yields. To achieve high sugar yields and relatively slower vegetative growth in the latter part of the growing period, progressively cooler nights should be accompanied by an exhaustion of available soil nitrogen and soil water. A seed crop requires low temperatures even up to 4°C for many weeks to induce flowering, which tends to be accelerated by long days.

But recently released varieties in India are suitable to be grown in **subtropical** regions and **tropical regions**. Their climatic requirements are vastly different from the temperate varieties. They can withstand higher temperatures (20–25°C for germination; 30–35°C for development and 25–30°C for sugar accumulation.

The crop can be grown on a wide range of soils with **medium** to **slightly heavy textured, well-drained soils** preferred. Soil pH smaller than 5.5 is unfavourable to growth. Crust forming at the soil surface, at the time of germination can lead to poor crop stand. Except during the early stages, the crop is **tolerant to salinity**. Yield decrease is 0% at EC of 7 m.mhos/cm, 10% at 8.7, 25% at 11, 50% at 15 and 100% at EC 24 m mhos/cm. However, during early growth, EC values should not exceed 3 m mhos/cm.

13.6 BOTANICAL DESCRIPTION OF PLANT

Sugar beet is a **biennial crop**, belonging to family *Chenopodiaceae*. However, it is grown and harvested as annual crop for sugar production, because the sugar content in roots is maximum at the end of first year's growth (Figure 13.1). Whenever, seeds are produced, second year growth is allowed, as flowering occurs during the second year.

The tap roots of sugar beet plant are well developed. The roots are tuberous, plumpy hard, whitish in colour, with high sucrose in them. Vertical grooves on opposite sides of tuberous roots are common. **Convex crown** of root (connecting to stem), middle broad **neck part** and lowest **long hairy lignified** structures are three parts found in the sugar beet roots. Maximum sugar is accumulated in the neck and crown. The stem is poorly developed with no branches (Figure 13.2). The leaves arise from the base of condensed stem in rosette shape. The leaves are simple, smooth and petiolate. Inflorescence (borne in second year) is terminal open branched panicle. Flowers are imperfect, regular and devoid of petals. Fruits are aggregate ball, with two or more seeds in each fruit.

Figure 13.1 Sugar beet crop during first grooved year with tuberous roots of sugar beet.
(*Images:* Courtesy Wikipedia)

Figure 13.2 Unbranched stem and no flower production.

13.7 GROWING PERIOD

The crop duration is extended from 140–200 days period. Long vegetative period, when the leaves accumulate sugar is followed by an extended post-vegetative period–when most of the sugar accumulates in the roots. The sugar yield is decided by both **root size** and its **sugar concentration**. With rapid growth of the storage root, the sugar concentration reaches a steady value which is principally determined by climate, water supply and nitrogen level in the soil and is influenced to some extent by variety and spacing. Time of sowing, affects the rate of crop development, particularly from emergence to peak height stage. For an autumn-sown crop the initial vegetative development stage may be 140 days, but for a spring-sown crop, it may not exceed about 90 days and for a late spring/early summer-sown crop the vegetative phase hardly exceeds 60 days. As the vegetative period is compressed, the size of the storage root as well as sugar synthesis is reduced.

13.8 VARIETIES AND HYBRIDS

LS-6 and IISR COMP-1 are two important varieties developed from IISR and its associates, recommended in many Northern states. Cauvery, Indus and Shubhra are the hybrids released and recommended for Tamil Nadu.

Similarities of Sugar Beet and Sugarcane
Both are sugar rich commercial crops, which are popular to produce sugar or ethanol along with number of by-products like bagasse and press mud (in sugarcane) molasses and large spent wash. Both have sugar in the high range of 15–21%. Both can be used to produce refined white sugar.

Differences between Sugar Beet and Sugarcane
Cogeneration of power is possible only in sugar industry based on sugarcane. But, sugarcane needs 11–15 months to mature, while sugar beet in India can mature in 5–7 months. Sugar beet is highly resistant to salinity than sugarcane. Water requirement of beet crop is smaller (70–85 cm), while sugarcane needs 150–170 cm.

13.9 CROPPING SYSTEMS

Being *rabi* crop, it is rotated with many *kharif* crops like maize, soybean, cow pea, black gram, green gram and even rice. Companion cropping sugar beet with autumn sugarcane is found to be very successful in improving sugar output per unit area.

13.10 NUTRIENT MANAGEMENT

Adequate nitrogen is required to ensure early maximum vegetative growth. Nitrogen is often given in split applications, one-fourth dose at planting and the rest in 2 splits–at thinning, and earthing up. An excessive dose of nitrogen or when applied late during the growing season reduces sugar content. Fertiliser applications of 120–150 kg/ha N, 50 to 70 kg/ha P at planting and 80–100 kg/ha K are necessary for the successful crop.

13.11 WATER MANAGEMENT

Total water requirements are in the range of 70–85 cm. But it can vary with climate and length of the total growing period. Sugar beet is particularly sensitive to water deficits at the time of **crop emergence** and **initial one month** after emergence. Frequent, light irrigations are preferred during this period, and irrigation may also be needed to reduce crust formation on the soil and to reduce salinity of the top soil. Except during emergence and early growth periods, it appears that the crop is less sensitive to moderate water deficits. The effect of reduced water supply over the total growing period is masked by influence of temperature, nitrogen availability and reduced water needs during a period just prior to the harvest. But, over-watering may retard leaf development and can encourage flowering during the first year (**bolting**). Sugar beet crop is a deep rooted crop and maximum water is extracted from top 0.8 to 1.1 m soil. Depletion of soil moisture can be allowed up to 50–60% of available soil moisture.

Light frequent irrigations are preferred in the initial stages, while soil moisture during thinning operation and harvesting operation is crucial. Except these stages, the irrigations can be scheduled in such a manner that soil depletion is allowed up to 50% of available soil moisture. At critical stages, however, more frequent irrigations are necessary.

13.12 WEED MANAGEMENT

Sugar beet crop is poor in resisting the growth of weeds up to 45 days, due to slow initial growth, after which the crop gains the ability to smother the weeds. 2–3 hand weeding or intercultivations are necessary to keep weeds under control. Herbicides like Pyramin (3 kg a.i./ha) or Betanal (2 kg a.i./ha) could be respectively used as pre-emergence spray or post-emergence spray (25–30 DAS) for easy control of the weeds.

13.13 PESTS AND DISEASES

Although, many diseases are reported in sugar beet, in many parts of the world, two important diseases are noticed in India. Three important insect pests are also noticed.

- **Sclerotium root rot:** White cottony growth along with deposit of mustard seed like structures are found on the roots in patches. The yields may be drastically reduced. To avoid the disease, mono cropping should be avoided. Excess N fertilisers or even excess irrigation may aggravate the disease. Drenching at the rate of 15 kg/ha during February can be adopted to prevent the spread of the disease.
- **Cercospora leaf spot:** Small circular ash coloured spots surrounded by yellowish/ reddish margin appear on the leaves. Its early occurrence may be dangerous to reduce the growth and yield.
- **Greasy cut worm:** Most common and widespread pest causing large scale damages in the initial stages of the crop. They cut the base of the plants during night and hide during day hours. It can be controlled by the spray of 5% Malathion dust @20 kg/ha.
- **Bihar hairy caterpillar:** A polyphagous pest feeds on leaf epidermis initially and skeletonises the leaves causing substantial damages to the growth. In the advanced stages, they may also make large holes. Insect can be controlled by dusting 2% methyl parathion dust @25 kg/ha.
- **Tobacco caterpillar:** In the initial stages, the damage appears to be similar to Bihar Hairy Caterpillar, but at later stages, the plants are completely defoliated. It can be controlled by the spray of carbaryl @1 kg per hectare in 1000L water.

13.14 OTHER PRODUCTION TECHNIQUES

- **Land preparation:** 1–2 deep ploughing followed by 2 harrowing and final levelling would prepare the land for sowing. **A finely pulverised smooth** and **levelled seed bed** is necessary. Before the final operation, 5–6 tons of organic manure is necessary to be applied.
- **Seeds and sowing:** The layout of the land depends on extent of levelling. In perfectly levelled lands, **flat bed** sowing can be undertaken, while ridges and furrow method is useful, when the levelling is not perfect. The rows of 45–50 cm are necessary to be established in either ways. The height of ridges could be maximum of 15–20 cm. The seeds could be dibbled and sown with drill at a depth not more than 1 inch. Deeper sowing can result into poor crop stand. Seed rate recommended for growing sugar beet is 8–10 kg/ha.
- **Season:** Sowing time is crucial for the success of the crop and may even decide the crop duration. Usually, sowing starts in the month of October and may extend up to November end. Early sowing is always advantageous in extending the sugar accumulation period.

- **After care:** Thinning after 15 days is crucial and essential operation to maintain optimum plant population.
- **Harvest and post-harvest:** Depending on the variety and its duration as well as sowing season, harvesting may start from March, but can get extended up to May. Invariably, the roots of sugar beet need to be loosened by tractor operated plough and later collected. The tops will have to be rejected and the sugar beets are expected to be sent to processing centre within 36 hours. A sugar beet yield of 30–35 tons/ha may be expected in India, although global average is as high as 50–55 tons/ha.

REFERENCES

Anonymous (2013), Sugar Beet Research and Development, IISR, Lucknow, pp. 5–20.

Anonymous (2013), Tropical Sugar Beet, TNAU Agri Tech Portal, pp. 5–8.

Anonymous (2014), FAOSTAT, http://faostat.fao.org/site/567/DesktopDefault.aspx?PageID=567 #ancor.

Anonymous, 2022, Agricultural statistics at a glance, DAFW, GoI, p. 29.

FAOSTAT, 2023, https://www.fao.org/faostat/en/#data/QCL.

Harveson, R.M. (2009), History of Sugar Beet Production and Its Use In: Crop Watch, University of Nebraska–Lincoln, http://cropwatch.unl.edu/sugarbeets/sugarbeet_history.

MODEL QUESTIONS AND KEY ANSWERS FOR RABI SUGAR CROPS

Choose Most Appropriate Answer and Fill in the Blanks

1. The initial set of roots arising in sugarcane is called .... (seminal roots/set roots/sett roots/crown roots)
2. Sugar beet contains ....% sucrose (20–24/12–21/10–12/8–13)
3. Accumulation of sucrose is higher in .... internodes of sugarcane(top/lower/middle/soft)
4. .... country is the largest producer of sugar in world (Pakistan/India/China/Brazil)
5. Nearly .... percent of sugarcane grown in India is used for jiggery (25/45/55/60)
6. Sugarcane matures in .... season (summer/winter/rainy/spring)
7. Sugarcane is propagated through the use of................ (hybrid seeds/certified seeds/stem cuttings/root cuttings)
8. Spring sown crop of sugar beet takes .... days for maturity (130–140/180–190/90–100/50–60)
9. The inflorescence of sugarcane is called .... (arrow/billow/spike/receme)
10. Total soluble solids in the cane juice is found out by ....values(EC/brix/Pol/LCC)
11. Sugar beet belongs to family .... (Euphorbiceae/Chenapodiaceae/Poaceae/Fabaceae)
12. Maximum sugar beet is grown in .... country of Asia (India/China/Turkey/Iran)
13. IISR is located in .... (Kanpur/Nagpur/Coimbatore/Lucknow)
14. Tying the tillers by dry leaves to reduce the lodging is called .... (propping/wrapping/ridging/polling)
15. A bio fuel produced from sugar beet is called .... (bioethanol/biobutanol/ biooctane/biodiesel)
16. The mass of true botanical seeds of sugarcane are called .... (puff/fluff/cuff/ruff)
17. Almost 1/5th of global sugar beet are is produced by .... (USA/China/Canada/Russia)
18. Conversion of sucrose into glucose is called .... (accretion/inversion/exversion/discretion)
19. Usually, a sugarcane sett contains not more than .... buds (2–3/3–5/4–7/1)
20. Top of the mature cane is used as planting material as it has .... (higher sucrose/hormones/active buds/higher moisture)
21. Pol in juice values for mature cane is in the range of ....% (19–24/11–16/20–25/15–20)
22. The ideal age of cane used as planting material is .... months (2–3/4–5/7–8/9–10)
23. Tropical sugar beet requires ....°C for better sugar accumulation (30–35/20–25/25–30/21–30)
24. .... found that sugar could be manufactured from sugar beet (Charles Morgan/Phillips Hadley/Andreas Margraff/Celestine Stanley)

25. The average recovery of jaggery from a ton of cane is .... kg (110–120/90–110/80–85/75–90)
26. Bud and chip method of planting sugarcane can .... (be more economical/establish good population/avoid diseases/induce hardness in seedlings)
27. Adsali cane is planted in the months of .... (Nov.–Dec./Jan.–Feb./June–July/April–May)
28. The begasse produced during crushing of cane is used to produce .... (fibre/molasses/power/cattle feed)
29. Seed crop of sugar beet is grown for a period of .... months (6–8/4–6/15–20/24–30)
30. Jaggery has................% of sucrose (55–60/98–99/80–88/75–78)

State TRUE or FALSE by Using 'T' or 'F'

1. Sugar beet is an example of stem modification. (T/F)
2. Sugarcane breeding Institute is located in Bangalore. (T/F)
3. Sugar accumulation in cane needs cold temperatures. (T/F)
4. Nearly 20% of the global sugar production is from sugar beet. (T/F)
5. Half of the sugar produced in India is from sugar beet. (T/F)
6. The sugar produced from sugar beet is brownish. (T/F)
7. Bora sugar is manufactured from local small scale units of Khandasari. (T/F)
8. The potassium fertilisers are applied to sugarcane in splits, due to its long duration. (T/F)
9. The average duration of planted cane is longer than ratoon cane. (T/F)
10. Red rot disease is transmitted through water. (T/F)
11. Cane sugar has higher sucrose than jaggery. (T/F)
12. Press mud, a product from sugar industry, is harmful to plants. (T/F)
13. Commercial cane sugar (CCS) indicates the purity percentage of cane. (T/F)
14. Molasses is produced from sugar beet and sugarcane. (T/F)
15. Grassy shoot is a type of tiller in sugarcane. (T/F)
16. Water shoots are desirable for higher yields. (T/F)
17. Biodiesel can be produced from molasses. (T/F)
18. More profuse branching helps beet root to produce more yields. (T/F)
19. Sugar beet is commercially grown as a biannual crop. (T/F)
20. Maximum purity percentage in sugarcane can be up to 35%. (T/F)

Answers

Fill in the blanks

1. Sett root	**11.** Chenapodiaceae	**21.** 11–16
2. 12–21 %	**12.** Turkey	**22.** 7–8
3. Lower	**13.** Lucknow	**23.** 25–30°C
4. Brazil	**14.** Propping	**24.** Andreas Marggraf
5. 45%	**15.** biomutanol	**25.** 90–110
6. winter	**16.** Fluff	**26.** establish good population
7. stem cuttings	**17.** Russia	**27.** June–July
8. 90–100	**18.** Inversion	**28.** power
9. Arrow	**19.** 2–3	**29.** 15–20 months
10. Brix	**20.** active buds	**30.** 75–78%

True or False

1. F	**2.** F	**3.** T	**4.** T	**5.** F	**6.** F	**7.** T	**8.** F	**9.** T	**10.** F
11. F	**12.** F	**13.** T	**14.** T	**15.** F	**16.** F	**17.** F	**18.** F	**19.** F	**20.** F

Part 5
MEDICINAL AND AROMATIC CROPS

Chapter 14

Citronella (*Cymbopogon winteratus*)

14.1 ECONOMIC IMPORTANCE

Citronella is one of the most important crops grown for aromatic oil. It is a popular commercial crop for many, as the farmers can easily afford to extract the **aromatic oil** and market them for good profit. Citronella grass is also known as **mosquito grass** due to its mosquito repelling character, or called **Java citronella**, due to its origin in Java Island. It has supported many industries like mosquito repellent industry, perfumery industry and incense stick industry, room freshener/sprayer industry besides many other sectors related to aromatic oils. But, its extraction into oil has not grown into a separate industry, because its fresh leaves should be crushed for extraction within 24-hour of harvesting, necessitating 'on the spot extraction' by the farmers themselves.

Citronella is often confused wrongly with *Citronella mucronata*, which is an ornamental plant in Chile and has no similarity with citronella grass—known for aromatic oil. An Australian rainforest tree, called *Citronella moorei* has no comparative character, but confused with citronella grass due to its name. Another plant called mosquito plant, scientifically called *Pelargonium citrosum* has no similarity with oil bearing 'citronella grass'.

14.2 ORIGIN AND HISTORY

Citronella oil yielding species *winteratus* was originated in **Indonesia** (**Java**), while species *nardus* was originated in **Sri Lanka**. Citronella was one of the world's **dominant insect repellents** before the introduction of DDT. Recent history has indicated that Citronella is once again becoming the product of choice for health conscious customers. Although, originated in Southern Asia, it is spread to USA, China and many other American and Asian countries in the last 60–80 years.

The oldest known records of using citronella oil and leaves as perfumes in religious ceremonies were found in India about 2000 years before. First evidence of using citronella oil in Sri Lanka has been reported by Dr. Nicolas Grim in the seventeenth century. By eighteenth century, Sri Lanka were a reputed exporter of citronella oil and the samples of Sri Lankan Citronella oil was reported to be displayed in World Trade auctions in London and Lisbon.

However, with the entering of Indonesia and a few other countries into market, the demand for Sri Lankan citronella has been declined. Demand for natural citronella has diminished to some extent, after the entry of **synthetic citronella** (out of petroleum distillation products), especially in perfumery industry.

14.3 CLASSIFICATION

Two species of citronella grass are identified. They are:

1. *Cymbopogon nardus* (Ceylon citronella), and
2. *Cymbopogon winteratus* (Java type citronella)

Like 16 other grasses belonging to *Cymbopogon* genus, the scented leaves and stems of these grasses are **bluish green** in colour, but containing volatile oil called **citronella oil.** Among the cultivated species of *Cymbopogon* genus, only *Cymbopogon nardus* and *Cymbopogon winteratus,* yield a characteristic 'citronella oil', although, other species also yield the aromatic oils. Ceylon citronella yields slightly inferior quality oil than Java citronella. Both species *nardus* and *winteratus* of genus *Cymbopogon* are **perennial**, **clumping grasses** and can grow up to 5–6 feet in height.

Both types probably originated from **Mana Grass,** occurs today in two wild forms—*Cymbopogon nardus* var. *linnae* (*typicus*) and *C. nardus* var. *confertiflorus*. Neither wild form is known to be used for **distillation** to any appreciable extent.

The higher proportions **of geraniol** and **citronellal** in the Java type make it a better source for perfumery derivatives. The name *Cymbopogon winteratus* is given to this selected variety to commemorate Mr. Winter—an important oil distiller of Ceylon, who first cultivated and distilled the Maha Pangeri type of citronella in Ceylon.

14.4 AREA AND DISTRIBUTION

At present, the world production of citronella oil is approximately 4,000 tons. The main producers are China and Indonesia producing 40% of the world's supply. The oil is also produced in Taiwan, Guatemala, Honduras, Brazil, Sri Lanka, India, Argentina, Ecuador, Jamaica, Madagascar, Mexico and South Africa. The market for natural citronella oil has been eroded by chemicals synthesised from turpentine derived from conifers. However, natural citronella oil and its derivatives are preferred by the perfume industry.

In India, the crop is grown in the states of Assam, Gujarat, Jammu & Kashmir, Karnataka, Maharashtra, Tamil Nadu, West Bengal and Uttar Pradesh.

14.5 SOIL AND CLIMATIC REQUIREMENTS

Citronella grass is best grown in **well-drained sandy loam soil** with abundant organic matter. **Heavy clay soils** and **sandy soils** do not support good growth of the plant. Variety of other soils, like medium black, red sandy, clay loam soils are also used to grown citronella. The plant has been found to grow well under a pH range of 5.8–6.0.

Although, 180–120 m altitude is optimum, the plants are reported to grow well at the altitudes between 1000–1500 m. Citronella thrives well under the **tropical** and **subtropical** conditions, with ideal temperatures varying from 24–30°C It requires abundant moisture and sunshine for good growth. A good rainfall of about 2000–2500 mm well spread over the year and high atmospheric humidity, appear to support good growth, yield and quality of the oil.

14.6 COMPOSITION OF CITRONELLA OIL

Citronella oil is a colourless or light yellow liquid with a characteristic **woody**, **grassy** or **lemony** odour. Citronella oil from *C. winteratus* are rich in citronellal, geraniol, geranyl acetate and limonene. But, oil from *C. nardus* consists of geraniol, limonene, citronellal, and methyl isoeugenol. The higher content of geraniol and citronellal in the Java type make it a better source for perfumery industry.

14.7 BOTANICAL DESCRIPTION OF PLANT

Citronella plant is tall growing, robust, profusely tillering grass (Figure 14.1) (height up to 2 m). It has soft glabrous soft, long and thin leaves 25–70/clump. The base of every tiller is **magenta** or **pinkish** colour and the whole plant smells **pleasant aroma** around it.

Figure 14.1 Profusely tillered Citronella grass.

14.8 VARIETIES

- **Mandakini:** Clonal selection, gives a little less herb yield (35 t/ha) and oils (118 kg/ha). The variety is suitable for hills and terai tracts of Himalayas (CIMAP, Lucknow).
- **Manjusha:** Clonal selection, gives an herbage yield of 43 t/ha and 150 kg/ha of oil per annum. The variety is suitable for Indo-Genetic plains (CIMAP, Lucknow).

- **Manjiri:** This variety has been released by the University of Agriculture Sciences, Bangalore. It is an elite mutant clone of Manjusha M 3–8. It has been found to yield 50–90% more oil (240–280 kg/ha/y), high citronellal and low geraniol content. It has profuse tillering and rapid growing ability, thus producing a high herb yields (upto 50 t/ha).
- **CIMAP Bio-13, Java-2, Jorhat-2:** High yielding varieties for the Southern and Eastern India released by CIMAP.

14.9 USES OF CITRONELLA OIL

- Citronella is credited with having therapeutic properties besides being used as antiseptic, deodorant, insecticide, tonic and as a stimulant.
- But, it is also more popular for its insecticide properties. Many commercial mosquito repellents contain citronella oil. Citronella oil can repel many insects like mosquito, ants, louses, flies. But, its use is absolutely safe to human beings.
- Diluted citronella oil, when applied to human skin, produce mild warmth, thereby relieves the muscular and joint pains. It is also found to be effective against migraine (severe persisting headache).
- Citronella can be popularly used as **trap crop** with many crops to reduce insect attack.
- It also used to manufacture soaps and candles.
- It has common applications in body massage. It is popularly used for treatment, called aroma therapy to patients suffering from tensions, stress, trauma, nervous fatigue, etc.
- This oil can also help with minor infection, but is more commonly known for its ability to assist in combating colds and flu. Citronella can also be used for excessive perspiration and for conditioning oily skin and hair.
- Its decoction in water is popularly consumed in place of tea or coffee, wrongly called citronella tea or gowti chai.

14.10 RISKS TO WILDLIFE, ANIMALS AND HUMAN BEINGS

If the citronella vapours are inhaled, it could cause an initial stimulation followed by the **depression** of the **central nervous system**. Citronella oil may be harmful if ingested in large quantity. Direct exposure to fumes may **irritate the skin and eye**. It has minimal or no risk to wildlife and environment, due to its toxic levels being low and its use being limited. The fumes of citronella oil are said to be potentially **toxic to birds**.

14.11 PRODUCTION TECHNIQUES

- **Land preparation and layout:** Thorough deep cultivation by using MB plough is necessary, followed by clod crushing and levelling. Later, the land is laid out in ridges and furrow after applying 8–10 tons of organic manure and mixing. The inter row distance is generally 90 cm for fertile and heavy soils and 60 cm for light and poor soils.

- **Planting:** Citronella is vegetatively propagated with culms. As it does not grow the runners, the base of culm itself split into slips (Figure 14.2), in such a way that each slip has at least 1–2 tillers retained. Slips are necessary to be prepared from existing plantation (1 acre plantation needs collection of slips from at least 4 guntha area from existing plantation). Each slip is planted at 60 cm spacing on ridges (18,000 plants/ ha @90 × 60 cm spacing) and irrigated within 1–2 days.

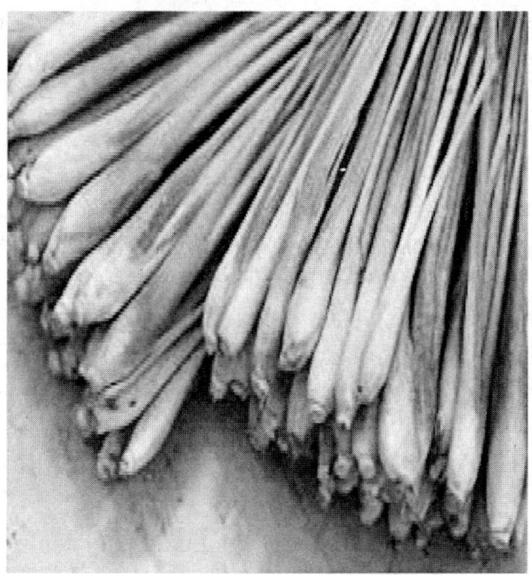

Figure 14.2 Slips of citronella for planting.

- **Planting Time:** Onset of monsoon or at the middle of rainy season i.e. June–July.
- **Fertilisers:** Nitrogen is applied @100 kg/ha in four splits (one-fourth at planting + one- fourth after 3 months + one-fourth after first cut + one-fourth after second cut. After all subsequent cuts, 25 kg N/ha will have to be supplied. A dose of 40 kg P_2O_5 and 40 kg K_2O per hectare should be given in single splits every year.
- **Improved Varieties:** Bio 13, Manjusha, Mandakini, Manjiri.
- **Weeding:** Weed control is necessary for the first 2–3 months till the slips are established, achieved by intercultivation or hand weeding. At later stages, crop will smother the weeds automatically.
- **Irrigation:** It requires regular irrigation, preferably once in 8 days in light soils or once in 12 days in heavy soils throughout the rainless period of the year. It could be irrigated by surface irrigation or drip irrigation method.
- **Pests and Diseases:** Citronella is affected by two important diseases—**leaf blight** and **anthracnose**. Leaf blight is controlled by the spray of Dithane M-45 or Z-78 (2–3 g/l). Anthracnose can be controlled by the application of Dithiocarbamate (1–2 ml/l).
- **Harvesting:** Three cuttings may be obtained in the first year commencing 6 months after planting. Leaf blades contain more oil than sheath, and hence harvesting is done 15–20 cm above the ground level. Optimum oil yield is obtained when the sixth leaf is well developed. After 1 year, the harvesting is done every two and a half months.

Citronella plantation can be harvested for 5 years without any reduced economic return. Later, it is necessary to replant the crop.

14.12 EXTRACTION OF CITRONELLA OIL

Fresh leaves should be used for extraction, and when they are allowed to be dried, the oil recovery is poor. These leaves are subjected to **hydro-distillation**, which involves boiling the leaves and collecting the citronella oil laden steam and condensing it to obtain the oil. Steam distillation can also be employed, wherein the steam is passed through the tissues, which rupture the tissues and carry the oil released in the process. Such steam is condensed to separate the oil. Generally, the citronella oil recovery is to the extent of 0.6–0.8% of the bio-mass. Hence, a huge quantity of biomass has to be subjected to extraction process to obtain small quantity of oil.

In recent years, vacuum technology or chemical extraction technologies have been developed to minimise the losses and improve the recovery. From farmers' perspective, a hydro-distillation involving a simple stainless steel extraction unit is less costly, easy to maintain and cost effective. However, the source of heat may be altered according to the convenience. Many improved versions of hydro-distillation are also developed.

REFERENCES

Anonymous (2009), Citronella–Soil and Climate, National Horticulture Board, http://nhb.gov.in/aroma/citronella/cit013.pdf.

Anonymous (2010), Citronella, Department of Export Agriculture, Government of Tamil Nadu http://www.exportagridept.gov.lk/web/index.php?option=com_content&view=article&id=129&Itemid=159&lang=en.

Anonymous (2010), Citronella Oil and its Uses, blogspot, http://citronella-oil.blogspot.in/ Anonymous (2011), Citronella Oil-Wikipedia, http://en.wikipedia.org/wiki/Citronella_http://www.edensgarden.com/citronella#.Uwous7vNsmwoil.

Anonymous (2012), Citronella Varieties, http://nhb.gov.in/aroma/citronella/cit014.pd.

Wany, Aakanksha, Shivesh Jha, Vinod Kumar Nigam and Dev Mani Pandey (2013), "Chemical Analysis and Therapeutic Uses of Citronella Oil from *Cymbopogon winteratus*—A Short Review", *International Journal of Advanced Research*, Vol. 1, Issue 6, pp. 504–521.

Chapter 15

Mentha (Mint) (*Mentha spp.*)

15.1 ECONOMIC IMPORTANCE

Mentha, also called **common mint** in English, is one of the most important aromatic and medicinal crops, having wide industrial application. Important species of mentha are known by several common names. In India, it is called **Pudina** and it occupies leading position in the world in respect of production and export of *mentha* oil. The word **mentha** is derived from Greek word called **Menthi**. The genus ***Mentha*** includes more than 18 species, of which four species are commercially important and remaining species are used for minor applications and few important are considered as invasive weeds. But, useful species are cultivated in all the continents. It entered India as late as 1960, as a result of its trial cultivation by M/s Richardson and Hindustan in the foot hill of Himalayas, for medicinal use of mentha oil.

15.2 ORIGIN AND HISTORY

Mentha was used in kitchens in most countries, for various culinary preparations, since time immemorial. *Mentha* was found in ancient Chinese literature and Egyptian civilisation. But, its importance as industrial crop was realised, only when Japanese started cultivation of *Mentha arvensis* on large scale and **extracting oil** from it since 1870. Different species of mint were originated in different parts of the world. The origin of most widely cultivated *Mentha arvensis* is obscure, because at least six variations of *Mentha arvensis* are identified and each of them was originated in different country. However, *M. arvensis* is said to be native of **China**, **Japan**, **Eurasia** and even **India** or **USA**. However, some authors hold that it was probably originated in **China**, but it was domesticated in Japan, initially spread to Europe and later spread to other part of world.

15.3 CLASSIFICATION

More than 18 species of mint are recognised, many of which have limited application and grown on a very limited scale or grown as wild. Four important species are commercially grown cool season crop.

These four important commercially important species of Mentha are recognised for diversified applications. They are:

1. **Japanese mint/Menthol mint** (*M. arvensis*) also called **Corn mint**, **Wild mint**, **Field mint**, **Pudina.**
2. **Pepper mint** (*M. piperita*) is a natural hybrid between water mint and spear mint.
3. **Spear mint** (*M. spicata*) also called ***M. virdis*** or ***M. cordifolia***, or curly mint.
4. **Bergamot mint** (*M. citrata*) also called **orange mint**.

Other types of mints are as follows:

• *Mentha aquatica*—Water mint, or Marsh mint • *Mentha asiatica*—Asian mint • *Mentha australis*—Australian mint • *Mentha canadensis* • *Mentha cervina*—Hart's Pennyroyal • *Mentha crispata*—Wrinkled-leaf mint • *Mentha dahurica*—Dahurian Thyme • *Mentha diemenica*—Slender mint	• *Mentha laxiflora*—Forest mint • *Mentha longifolia*—*Mentha sylvestris*, Horse mint • *Mentha pulegium*—Pennyroyal • *Mentha requienii*—Corsican mint • *Mentha sachalinensis*—Garden mint • *Mentha satureioides*—Native Pennyroyal • *Mentha suaveolens*—Apple mint, Pineapple mint (a variegated cultivar of Apple mint) • *Mentha vagans*—Gray mint

Note: *M. longifolia* and *M. suaveolens* cover the soil and has anti-eroding property; concentrated oil of *M. pulegium* is poisonous; *M. requieni* is shade tolerant.

15.4 AREA AND DISTRIBUTION

As all four commercially grown four species have differing composition of their respective oils, and these oils have equally diversified applications, different countries known for *mentha* cultivation grow all four species. In India, Japanese mint, spear mint and pepper mint are grown on large scale and their oils are being exported. USA is a leading producer of spear mint and pepper mint. *Mentha* cultivation and production of *mentha* oil are mainly taken up in India (largest producer and exporter), China, Brazil and USA. India produces 12,000 tons of menthol, out of global production of 16,000 tons. Other countries growing different types of *mentha* are Paraguay, Argentina, Japan, Thailand, Angola, Morocco, Argentina, Australia, France, USSR, Bulgaria, Czechoslovakia, Hungary, Italy, Switzerland and on a small scale in many European countries.

In India, area under cultivation of *mentha* are found in Uttar Pradesh (56% of Indian area), Punjab, Haryana and on minor scale, in the states of Madhya Pradesh and Gujarat. It is estimated that Indian area under *mentha* is around 2.25 lakh hectare, with Uttar Pradesh alone contributing to 1,25,000 hectare.

Position of India in World Market for Mentha Oil
India exports different types of mint oils to a number of countries including Argentina, Brazil, France, Germany, Japan, UK, USA, etc. These varieties include the Japanese mint oil (derived from *Mentha arvensis*), peppermint oil (*Mentha piperita*), dementholised Japanese mint oil, spear mint oil (*Mentha spicata*), water mint oil (*Mentha aquatica*), horse mint oil (*Mentha sylvestris*), Bergamont oil (*Mentha citrata*) and still others. India produces 12,000 tons of Japanese mint oil out of 16,000 tons of the global production. In respect of Bergamont oil, India produces 150 tons out of the global production of 200 tons. But, Indian contribution in respect of Pepper mint (5%) and spear mint (15%) is relatively lesser out of total global production.

15.5 SOIL AND CLIMATIC REQUIREMENTS

Japanese mint can be grown in **tropical** and **subtropical** climates with irrigation. It cannot tolerate wet winters (due to root rot). A temperature of 20–25°C promotes vegetative growth, but the essential oil and menthol are reported to increase at a higher temperature of 30°C under Indian conditions.

Pepper mint and spear mint cannot be grown profitably in tropical and subtropical areas, especially with very high summer temperature. They are ideally grown in **humid** and **temperate** conditions like Kashmir and foot hills Himalaya of Uttar Pradesh and Himachal Pradesh.

Bergamot mint can be grown in **temperate** and **subtropical** climates, however, its yields are higher in temperate climates. In general, open sunny climate with well distributed light rains during the growing period is congenial for good growth and development of the oil.

Medium to fertile deep soil, rich in **humus** is ideal. The soil should have good water holding capacity, but water logging should be avoided. A pH range of 6 –7.5 is ideal.

15.6 BOTANICAL DESCRIPTION OF PLANT

Mint belongs to family *Lamiaceae* and is a aromatic herb, grown as **perennial** or **annual** plant. They have wide-spreading underground runners and over ground *stolons* and erect, square, branched stems. The leaves are arranged in opposite pairs, from oblong to *lanceolate*, often fleshy (Figure 15.1), and with a serrated margin. Leaf colours range from dark green and grey-green to purple, blue, and sometimes pale yellow. The flowers are white to purple and produced in false whorls called **verticillasters** (Figure 15.2). The corolla is two-lipped with four sub equal lobes, the upper lobe usually the largest. The fruit is a nut let, containing one to four seeds.

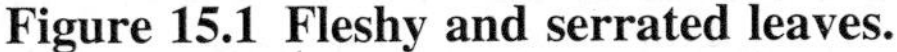

Figure 15.1 Fleshy and serrated leaves.

Figure 15.2 Verticillaster of mint.

While the species that make up the Mentha genus are widely distributed and can be found in many environments, it is best grown in humid climates. Mints will grow 10–120 cm tall and can spread over an indeterminate area. Due to their tendency to spread unchecked, some species of mints are considered invasive weeds.

15.7 VARIETIES AND HYBRIDS

- **Japanese mint:** CIMAP-MAS-1 and CIMAP-Hybrid-77, Shivalik, EC-41911, Gombi, Himalaya, Kalka, Kosi, Gomati, Damroo, Sambhav and Saksham Spear mint: CIMAP-MSS-1, CIMAP-MSS-5 and CIMAP-MSS-98, Punjab spearmint-1, Ganga, Neerkalka
- **Bergamot mint:** Kiran
- **Peppermint:** Kukrail, Pranjal, Tushar

15.8 SEASONS

In the subtropical plains, planting is done during the **winter** months, whereas in temperate climates, planting is done in **autumn** or **spring** from the last week of December to the first week of March or from the first week of January to the third week of February. Late planting always gives poor yields.

15.9 CHEMICAL COMPOSITION OF MINT OIL

- ***Japanese mint* (*M. arvensis*):** The fresh leaves of Japanese mint contain 0.4–0.6% oil. The oil contains menthol (65–75%), menthone (7–10%) and menthyl acetate (12–15%) and terpenes (pipene, limonene and comphene).
- ***Peppermint* (*M. piperita*):** The fresh leaves of peppermint contain 0.4 to 0.6% essential oils. Its constituents are almost similar to Japanese mint oil, with the exception of menthol, which is lower in peppermint (35–50%). The other constituents are menthyl acetate (14–15%), menthone (9%) menthoufuran and terpenes like pinene and limonene.
- ***Bergamot mint* (*M. citrata*):** Linalool and linalyl acetate are the main constituents of Bergamot mint oil. The oil is used directly in perfumes.
- ***Spearmint* (*M. spicata*):** The major constituent of spearmint oil is carvone (57.71%) and the other minor constituents are phellandrene, limonene, L-pinene and cinelole.

15.10 USES OF MINT OIL

Japanese mint oil (also called **menthol mint oil**) is mainly used to manufacture a commercial product—menthol, which is used widely to manufacture toothpastes, mouth washes, chewing gums and candies. Japanese mint oil is also used to manufacture dementholised oil (DMO), which has constituents similar to peppermint oil.

Menthol mint oil is also used to enhance shelf life of food grains and other edible products. It is also a very popular ingredient to manufacture ointments, pain balms, lozenges, syrups, deodorants, colognes, and cosmetics.

The spearmint oil is used mostly for flavouring in the toothpastes and as food flavouring in pickles and spices, chewing gum and confectionery, soaps and sauces, while bergamot mint oil is used in the cosmetic preparations like scents, soaps, after-shave lotions and colognes.

More recently, menthol mint oil is being used as anti-fungal agent, insect repellent, and fumigant. DMO has been found to have nitrification inhibitory properties and attempted to be used as coating over urea to increase N use efficiency.

15.11 PROPAGATION

Stolons and **suckers** are popularly used to propagate mint crop. Even **runners** are used to propagate peppermint and bergamot mint. The stolons or runners are cut in smaller pieces of around 10 cm and planted in ridges at a depth of 5–7 cm and irrigated immediately. They are prepared from, the existing field (approximately one-tenth of the planted area is needed to collect the stolons or suckers or runners) during the month of December or January. Around 400 kg of planting material is needed for 1 ha to plant them @ spacing of 45–60 cm rows.

A nursery also could be established in a bed and after 30–35 days, the seedlings are transplanted in the main field–instead of direct planting by stolons.

15.12 CROPPING SYSTEMS

Mint occupies the land for 180 –200 days of a year, and generally fits into 1-year sequences like mint–maize/potato or mint–paddy/potato. Two tear sequences like mint–wheat–paddy, mint–fallow–fodder millet are also practiced in parts of Punjab.

15.13 NUTRIENT MANAGEMENT

Mint is a **nitro positive** crop, with excellent response to applied nitrogen, which is species dependent. *M. arvensis* varieties respond up to 160 kg N/ha, while for *M. piperate* and *M. citrate,* the response up to 125 kg N/ha and for *M. spicata*, the maximum response was observed up to 120 kg/ha. In general, it is useful to apply nitrogen in three splits, i.e., after 45 DAS, after 90 DAS and third after first harvest. The phosphorus and potassium doses required by the crop are in the range of 50–60 kg/ha and 30–40 kg/ha respectively. Application of manure at the time of planting @4–5 t/ha is useful.

Mint also respond well to micronutrient applications like 20 kg $ZnSO_4$/ha and 20 kg sulphur/ha, applied at the time of planting, wherever the soils are deficient in these nutrients.

15.14 WATER MANAGEMENT

In heavy clay soils, with higher water holding capacity, the frequency of irrigation can be once in 12–15 days during rainless period and in lighter soils, including loamy soils, it can once in 8–10 days. Its water requirement is as high as 80–85 cm.

15.15 WEED MANAGEMENT

Critical competition by weeds is extended up to 70 days. Regular intercultivations can help to control the weeds effectively, duly along with a hand weeding, depending upon the weed intensity. In plots with high weed density history—it is useful to apply pre-emergence application of 1 kg/ha Oxyfluorfen or post-emergence of 2.5 l Gramoxone/ha (directed spray) or 4 kg/ha Dalapon.

15.16 HARVESTING AND YIELD

Mentha is harvested by cutting **entire foliage** leaving 10 cm above the soil after 100–120 days after planting, followed by a subsequent harvest after 180–200 days. The foliage yield may range from 20–25 t/ha. in a normal crop, while a good crop may yield up to 40 t/ha. The average oil recovery is around 0.4%, resulting into an oil yield of 75–80 kg/ha.

15.17 PESTS AND DISEASES

- **Stolon rot:** A serious disease of *M. arvensis* causing heavy loss. The seedlings fail to establish by rot of stolons. It can be controlled by Mancozeb or Dithane–M–45 @4 g/l.
- **Powdery mildew:** Starts as white patches on leaves, which later spread to other parts of the plant. These plants show leaf droppage after yellowing. The disease is controlled by any copper fungicide like Blitox.
- **Leaf blight:** Leaves show dark brown spots with concentric rings surrounded by pale yellow margins. The leaves are dropped later. It can be controlled by the use of Chlorothalonil.
- **Leaf rust:** They are characterised by small brown spots on leaves, causing heavy loss of leaf yield by disease induced defoliation. It can be controlled by the use of wettable sulphur or copper fungicide, before the damage is observed.
- **Verticillum wilt:** It is one of the most serious diseases of mint. The attacked plants fail to develop adequate foliage and leaves develop bronze colour and later wither away. It is a soil-borne disease. No perfect control measure is found against this disease. However, use of good seed material, rotation with non-host crops or use of 1, 3–Dichloropropene have offered partial control of this disease.
- **Cut worm:** It is a serious pest of the crop. It cuts the young plants at the base and brings out patchy field. It is controlled by the spray of Metasystox @1 ml/l.
- **Bihar hairy caterpillar and semi looper:** The caterpillars can defoliate the leaves causing serious loss of herbage yield. It is controlled by the spray of Methyl parathion @2 ml/l or Endosulphan @1.5 ml/l.
- **Mentha leaf roller:** The leaves roll and caterpillar thrives inside the roll, causing serious yield reduction. It is controlled by the spray of cypermethrin @½ ml/l.
- **White fly:** Sucking the sap from leaves will reduce the vitality of leaves, which are later dropped. It is controlled by the spray of rosin or nemazol @1 ml/l.

15.18 PROCESSING THE MINT OIL

Mint oil is distilled by using **fresh herbage** or after drying 1–2 days. In either case, the stems will have to be removed as they contain very little oil. The distillation can be by passing steam through the herbage and condensing the steam later. Conventional **hydro distillation** is also employed by many farmers.

REFERENCES

Anonymous (2010), Mint, National Horticultural Board http://nhb.gov.in/Horticulture%20 Crops%5CMint%5CMint1.htm.

Anonymous (2011), Brief report on mentha oil. Pub: Ventura Commodity, pp. 1–9.

Anonymous (2011), Mentha–Lecture Series, TNAU, Coimbatore, pp. 1–7. Anonymous (2012), *Mentha* Oil, CNR India, pp. 1–25.

Anonymous (2012), Mints, www://indg.in.

Anonymous (2013), *Mentha arvensis*–Wikipedia, en.wikipedia.org/wiki/Mentha_arvensis.

Anonymous (2013), Mentha, http://en.wikipedia.org/wiki/Mentha.

Chand, Sukhmal, Patra, N.K., Anwar, N. and Patra, D.D. (2004), Agronomy and Uses of Menthol Mint–Indian Perspective Procedings, Indian National Science Academy, B70, No. 3, pp. 269–297.

Rokosz, Maria Kostka, Yelena Yalli, Lana Dvorkin Pharm D., Julia Whelan, M.S. (2008), *Mentha arvensis piperascens,* Boston University of School of Medicine.

Chapter 16

Lemon Grass (*Cymbopogon citratus* and *Cymbopogon flexuosus* L.)

16.1 ECONOMIC IMPORTANCE

Lemon grass is an important commercial crop for aromatic oil, with diversified applications and is in greater demand world over. It is a perennial grass, resembling citronella grass morphologically, but differs from it in many respects. Two species, namely *citratus* and *flexuosus* are considered as **true lemon grasses** (due to commercial viability), although, all the members of genus *Cymbopogon* are sometimes wrongly referred as lemon grass. The lemon grass oil is rich in **citral** which has large number of applications in the fields of pharmaceuticals, perfumery, cosmetics, etc.

16.2 ORIGIN AND HISTORY

Lemon grass has two distinct species namely *citratus* (called **East Indian/Malbar/Cochin lemon grass**) and *flexuosus* (called **West Indian lemon grass**). The former species was originated in **Indo-Burmese region** and domesticated in India, Burma, Thailand and Sri Lanka. The latter species was originated in **Malaysian region** and domesticated in South India, Sri Lanka, Indonesia and Malaysia. Although, it was originated in Asia, it is now spread largely to Africa, besides different countries of Asia, due to its highly diversified industrial applications.

16.3 AREA AND DISTRIBUTION

Globally, lemon grass is mainly distributed in Africa and Asia. India is the largest producer (300 tons/year) followed by Guatemala in Africa. Other countries producing lemon grass are China, Zambia, Sri Lanka and Madagascar. India produced and exported as high as 1800 tons oil a year earlier. But, many other countries have entered fray and Indian area has reduced to 3000 hectare. China exports lemon grass oil as well as oil from *Litsea cubeba* rich in citral, thus restricting the export market for other countries (Lemon grass oil is also rich in citral). In India, it is mainly produced in heavy rainfall hilly tracts of Western Ghats in Kerala, Karnataka and Maharashtra as well as foot hills of Uttar Pradesh, Assam and Arunachal Pradesh.

16.4 BOTANICAL DESCRIPTION OF PLANT

Lemon grass is a tall, perennial fast growing grass with tuft of **lemon scented leaves** from a short rhizome (Figure 16.1). Leaves are long, glaucous, green, linear, tapering upwards and along the margins; ligule very short; sheaths tightly clasp at the base, but narrow and separating at distal end. It is a **short day plant** and flowers profusely in South India. The inflorescence is a long spike about one metre in length. Panicles are 30 to over 60 cm long.

Figure 16.1 Tuft of leaves of lemon grass.

Cymbopogon flexuosus grows to a height of 1.5 m, with distinct, dark-green foliage and produces seed. *Cymbopogon citratus* grows to a height of 1 m. It has distinct **bluish-green leaves** and usually does not produce seed. Both these grasses produce many bulbous stems that increase the clump diameter as the plants mature.

16.5 SOIL AND CLIMATIC REQUIREMENTS

Ideal conditions for growing lemon grass are **warm** and **humid climate** with well distributed rainfall of 250–330 cm per annum, both in tropical or subtropical climates. A temperature ranging from 20–30°C and **good sunshine** throughout the year is conducive to high crop yield with better oil content. For this reason, in subtropics, the oil content is likely to be reduced, especially, where the crop experiences cold temperatures below 12°C. In the hilly areas receiving heavy rainfall, the plant grows luxuriantly and is harvested more frequently, but the oil and citral content are less as compared to the plants growing in the regions of less rainfall. Lemon grass can also be grown in the semi-arid regions receiving low to moderate rainfall.

Cymbopogon flexuosus grows best at an altitude of up to 2200 m above the sea level.

Cymbopogon citratus grows best at an altitude of up to 750 m above the sea level.

It flourishes on a wide variety of soils ranging from loam to poor laterite, calcareous soils. However, well drained, sandy-loam soils are ideally suited for better growth, yield and oil content. Soils with poor drainage and with prolonged water logging should be avoided.

16.6 VARIETIES

Important varieties of lemon grass grown in India are presented in Table 16.1.

Variety	*Description*
Sugandhi (OD 19)	• It is adapted to a wide range of soil and climatic condition. • A red stemmed variety with plant height 1 to 1.75 m and profuse tillering. • The oil yield ranges from 80 to 100 kg per hectare with 85–88% of total itral produced under rainfed conditions (with life-saving irrigation).
Pragati	• It is a tall growing variety with dark purple leaf sheath suitable for North Indian plains and tarai belt of subtropical and tropical climate. • Average oil content is 0.63% with 75–82% citral.
Praman	• It is a medium sized variety with erect leaves and profuse tillering. • The oil yield is high with 82% citral.
Jama Rosa	• Very hardy with vigorous growth. • The variety yields about 35 tons of herbage per hectare containing 0.4% oil (FWB). • The variety yields up to 300 kg oil in 4–5 cuts in 16–18 months growing period.
RRL 16	• Average herbage yield of this variety is 15 to 20 tons/hectare/annum giving 100 to 110 kg oil. • The oil content varies from 0.6 to 0.8% (fresh weight basis) with 80% citral.
CKP 25	• A hybrid between *C. khasianum* × *C. pendulus*. • Gives 60 t/ha herbage in North Indian plains under irrigation. • The oil contains 82–85% citral.
Other Varieties	• OD-408, Kaveri • Kaveri needs high soil moisture to produce luxuriant growth and is evolved for the river valley tracts. Its characters are similar to those of OD 19.

16.7 USES OF LEMON GRASS OIL

- The oil has strong lemon-like odour, due to high percentage (over 75%) of citral in the oil. The characteristic smell of oil makes its use in scenting of soaps, detergents and insect repellent preparations.
- Major use of oil is in perfumery, cosmetics, beverages and is useful to manufacture ionones, which produces vitamin A.
- Citral rich oil has germicidal, medicinal and flavouring properties. Hence, it is used in many antiseptics, and culinary purposes.
- West Indian lemon grass (*C. citratus*) is low in citral content in the oil and has meager trade in the country.

16.8 COMPOSITION OF OIL

Lemon grass oil is a clear pale yellow liquid with a strong lemon odour with relative gravity of 0.87–0.90. The oil from *Cymbopogon citratus* contains 67% citral and 7% geraniol and negligible citronella oil. But, oil from *Cymbopogon flexuosus* has 21% citral and 62% geraniol and 2% citronella oil. Hence, the utility of these oils in different applications are grossly different.

16.9 AGRO TECHNIQUES TO CULTIVATE LEMON GRASS

- **Land preparation:** The land has to be ploughed twice, harrowed, levelled and laid out into ridges and furrow layout, after applying organic manure @8–10 t/ha.
- **Planting:** For better quality oil and higher oil yield, it is suggested to grow lemon grass by slips (Figure 16.2) obtained by dividing well-grown clumps. The lower brown sheath should be removed to expose young roots. However, seeds could be used to establish a crop of *C. flexuosus.*

Figure 16.2 Slips of lemon grass.

- **Harvesting and preparation of seed:** The crop flowers during November–December and seeds mature in next two months, viz. February–March. Seeds are collected from such healthy plants, which are not harvested for foliage. On an average, a healthy plant gives about 100–200 g of seeds. At the time of seed collection, the whole inflorescence is cut and sun dried for 2–3 days. These are then threshed and seeds are again dried in the sun and the seed remain attached with fluffy mass which is removed by beating of under the bag. These dry seed lots are stored in gunny bags lined with polythene. The seeds lose their viability, if stored for a period more than one year.
- **Raising the nursery:** The seeds are sown by hand on well prepared and manured (20 beds are needed per hectare) raised beds of 1 m–1.5 m width at the onset of monsoon and are covered with a thin layer of soil, using a seed rate of 4–5 kg/ha. The

bed should be watered immediately after sowing and care should be taken to maintain adequate moisture in the soil. Seeds germinate in 5–6 days and the seedlings are ready for transplanting after a period of 60 days.

- **Population density:** Irrespective of using the slips or seedlings raised in nursery, a spacing of 40 × 40 cm (62,500 plants/ha) or a spacing of 60 × 40 cm (41,600 plants per hectare) is necessary in a high rainfall area or under irrigation.
- **Planting technique:** Planting of slips is necessary in the holes of 10 –15 cm depth. An initial high planting rate can be used and as the plants mature, every second plant can be taken out and divided again for new slips. The space created in this way will be filled with remaining plants as they mature. Planting of slips or seedlings can be done during June–July. It is necessary to avoid planting during very hot times of the year and during winter.
- **Fertiliser management:** Looking to large biomass harvested, it is recommended to use a dose of fertilisers supplying N, P_2O_5 and K_2O @150:60:60 kg/ha/year. Application of 30 kg nitrogen, and entire P_2O_5 and K_2O per hectare as basal dose at the time of planting is necessary. The remaining nitrogen can be applied as top-dressing in 3 to 4 split doses during the growing season. In poorer soils, the dose of nitrogen should be increased. In zinc deficient soils, application of zinc sulphate @25 to 50 kg per hectare is recommended. avoided as the plants may develop root-rot during the rainy season.
- **Irrigation:** The improved varieties of lemon grass perform well with supplemental irrigation. Depending upon the rainfall and its distribution, the field is to be irrigated at an interval of 3 days during the first month and 7–10 day intervals subsequently. After the establishment of plants, irrigation schedule is adjusted depending on the water holding capacity of the soil and weather conditions.
- **Weed control:** The field is kept free of weeds for the first 3–4 months after plating. Similarly, weeding-cum hoeing is done up to 1 month, after every harvest. Generally, 2–3 weeding are necessary during a year. In row-planted crops, intercultural operations can be done by a tractor-drawn cultivator or hand-hoe. Application of distillation waste from the crop as organic mulch @3 MT/ha will help in controlling weeds and retaining soil moisture. Chemical weed control through application of Diuron @1.5 kg a.i./ha and Oxyfluorfen @0.5 kg a.i./ha are effective for weed control.
- **Harvesting:** The first harvest is generally obtained after 4 to 6 months of transplanting seedlings. Subsequent harvests are done at intervals of 60–70 days depending upon the fertility of the soil and other seasonal factors. Under normal conditions, three harvests are possible during the first year, and four harvests in subsequent years, depending on the management practices followed. Harvesting is done with the help of sickles, the plants are cut 10 cm above ground-level and allowed to wilt in the field for a day or two, before transporting to the distillation site.
- **Age of plantation and recovery:** Depending upon soil and climatic conditions, the plantation lasts on an average, for 4 to 5 years. The yield of oil is less during the first year, but it increases in the second year and reaches a maximum in the third year; after this, the yield declines. On an average, 20 to 30 tons of fresh herbage is harvested per hectare per annum from 3–4 cuttings. The oil yield varies from 0.5% to 0.8% depending on the variety, season/month of harvest and age of the crop, with an average oil yield of 0.65%.

- **Pest management:** Stem borer is most important pest, which can attack the crop. The caterpillar is white in colour with a black head and black spots on the body. It bores into the stem and remains there, feeding on the shoot. It is usually found at the bottom of the stem. The first symptom of the attack is the drying up of the central leaf. Subsequently, the entire shoot dies, resulting in a significant reduction in biomass yield and oil yield.

16.10 OIL EXTRACTION PROCESS

The oil is mainly obtained through steam distillation of the **dried material**. The recovery of oil from local dried lemon grass is about 0.7%. Hydrosol (floral water), a by-product of the process, can be used in cosmetic, toiletries and herbal products. Zero waste processing can be realised as the spent leaves can be used in personal care products, animal feed and compost. Purification of the oil can be achieved by simple filtration. Extraction using the spinning cone column employs a somewhat more sophisticated technology. It is thought to be a very efficient extraction process which can be manipulated to produce specific profiles.

REFERENCES

Anonymous (2009), Lemon Grass, *Information Bulletin*, Scientific Research Council, Jamaica, www.src-jamaica.org.

Anonymous (2009), *Lemon Grass Production*, Department of Forestry and Fisheries, Republic of South Africa, pp. 2–26.

Anonymous (2010), *Cultivation and Processing of Lemon Grass*, Department of Agriculture, Government of Puducherry, India, pp. 2–15.

Anonymous (2011), *Lemon Grass*, National Horticultural Board, Government of India, pp. 1–14, ttp://nhb.gov.in/Horticulture%20Crops%5CLemongrass%5CLemongrass1.htm.

MODEL QUESTIONS AND KEY ANSWERS FOR MEDICINAL AND AROMATIC CROPS

Choose Most Appropriate Answer and Fill in the Blanks

1. The term 'mosquito grass' denotes .... (citronella/lemon grass/a mosquito weed)
2. Higher fraction of geraniol and citronellal make .... type of citronella grass suitable for perfumery industry. (Java/Lanka/Ceylon/Sumatra)
3. The scientific name of true lemon grass is .... (*Cymbopogon nardus/Cymbopogon citratus/Cymbopogon martini/Cymbopogon winterata*)
4. A type of tea can be made from .... grass. (lemon/citronella/ *mentha*)
5. The crop, which is repellent against many insects, is also attractant of honey bees .... (lemon grass/citronella grass/palmarosa/scented fodder)
6. Menthol mint refers to *Mentha* .... *(asiatica/aquatic/arvensis/vagans)*
7. Majority of *mentha* oil production takes place in the state of .... (Madhya Pradesh/ Gujarat/Karnataka/Uttar Pradesh)
8. National Institute for Medicinal and Aromatic plants is located in .... (Bangalore/ Lucknow/Anand/Vishakhapatnam)
9. 'Mandakini' is the name of variety of .... (citronella/lemon grass/*mentha*)
10. Recovery of oil from citronella grass is .... (1–1.5%/8–10%, 0.6–0.8%/0.03–0.08%)
11. The average foliage yields of *mentha* crop ranges between............... (20–25 t/ha/5–10 t/ ha/0.8–1.5 t/ha/4–5 t/ha).
12. Most of the aromatic oils are extracted by the process called .... (acid conversion/ carbon distillation/hydro distillation/alkaline phosphatation)
13. Lemon like smell in lemon grass is due to .... (geranoil/citral/menthol/citronellal)
14. .... crop is called pudina in India. (*Mentha*/Palmarosa/Citronella/lemon grass)
15. Sugandhi is the name of variety of .... (*mentha*/palmarosa/citronella/lemon grass)
16. Central Institute of Medicinal and Aromatic Plants (CIMAP) is located in .... (Jabalpur/Ernakulum/Bangalore/Srinagar)

State TRUE or FALSE by Using 'T' or 'F'

1. The average oil yield from *mentha* crop is around 9–10 q/ha. (T/F)
2. *Mentha* belongs to family *Lamiaceae.* (T/F)
3. Mint crop is propagated by stolons or suckers. (T/F)
4. Majority of area under mint in India is found in Uttar Pradesh. (T/F)
5. USA mainly produces pepper mint. (T/F)
6. Lemon grass is grown mainly in hilly regions of Kashmir and Uttarakhand. (T/F)
7. Psyllium is the name given to lemon grass. (T/F)
8. Scientific name of citronella grass is *Cymbopogon winterianus.* (T/F)

9. Bergamot mint's scientific name is *Mentha citrata.* (T/F)
10. West Indian lemon grass is same as *Cymbopogon flexuosus.* (T/F)
11. East Indian lemon grass was originated in China. (T/F)
12. Menthol mint is originated in China. (T/F)
13. Indian produces 40% of the global citronella oil. (T/F)
14. Citronella oil is mainly consumed in perfumery industry. (T/F)
15. Central Institute for Aromatic and Medicinal Plants is located in Srinagar. (T/F)

Answers

Fill in the blanks

1. citronella
2. Java
3. C. citratus
4. Lemon
5. lemon grass
6. arvensis
7. Uttar Pradesh
8. Anand
9. citronella
10. 0.6–0.8%
11. 20–25 t/ha
12. Hydrodistillation
13. citral
14. Mentha
15. llemon grass
16. Bangalore

True or False

1. F **2.** T **3.** T **4.** T **5.** T **6.** F **7.** F **8.** T **9.** T **10.** T
11. F **12.** T **13.** F **14.** T **15.** F

Part 6
COMMERCIAL CROPS

Chapter 17

Potato (*Solanum tuberosum* L.)

17.1 ECONOMIC IMPORTANCE

Potato is an important food crop of the world, although, it is also classified as a **commercial crop**. It is fourth important food crop of the world after rice, wheat and maize. Its ability to yield maximum starch output per unit time has held it as the fastest starch producing crop. Dietary energy output of potato is as high as 77 kcal per 100 g of fresh potato. Even in the terms of commercial yield per unit area and per unit time, potato ranks first, as no other food crop has such a fast output. It can yield 20–30 tons per hectare (2–3 kg/m^2) in 130 days (154–230 kg per day per hectare).

Besides being used as a vegetable throughout the world, it is known for the production of starch, fermented products, ready to eat food articles (like chips). It has been an integral part of the human food, within four centuries of its introduction from its origin to different countries of the world. Potato with skin contains around 17% carbohydrate (90% of which is starch), 2% protein and 0.1% fat. Unpeeled potato contains considerable quantity of B_3, B_5 and B_6 and Vitamin C, besides minerals such as potassium, phosphorus and magnesium. However, it is poor in iron, zinc and manganese and other vitamins.

Out of total starch, around 7% of the harvested tuber weight (out of 17% carbohydrate on weight basis) is resistant to digestion in stomach and small intestine and reaches large intestine undigested. Such a portion serves as dietary fibre. Fraction of crude fibre gets increased to 13% from 7%, once the potatoes are cooked and cooled. For this reason, it is popularly known to reduce the incidence of colon cancer, to improve the glucose tolerance as well as insulin sensitivity.

Potato production in the world is restricted to limited areas, but its consumption is throughout the world. This feature is found even in India. Similarly, potato is grown in specific season, but consumed throughout the year. This feature calls for comprehensive storage and transport facilities to supply potato in the regions not growing it or at time, when it is not grown.

Due to its **short duration**, **shrubby**, and **uncompetitive nature of growth**, it is popular both in **sequential** and **intercropping systems**—both under irrigation and rainfed situations.

17.2 ORIGIN AND HISTORY

Potato was originated between **Peru** and **North Western Bolivia** in **Andes hill** of **South America** from wild form called *Solanum brevicaule* L. It was first domesticated in Peru about 7000–10,000 years back. By repeated selective breeding, several subspecies were developed, but most of the modern varieties were developed from **Chilean subspecies**.

Spanish introduced potato to Europeans during sixteenth century and European mariners spread throughout the world, wherever they spread their empires. In India and China, who are now major producers of potato, it was introduced respectively during eighteenth and nineteenth century. Most of the South and North American countries as well as European countries grow it as staple food, although, in the last 50–60 years, the centre of production to South Asian countries.

17.3 CLASSIFICATION

Major species grown worldwide is tetraploid *Solanum tuberosum* ($2n = 48$), and modern varieties of this species are the most widely cultivated. There are also four diploid species ($2n = 24$): *S. stenotomum*, *S. phureja*, *S. goniocalyx*, and *S. ajanhuiri*. There are two triploid species ($3n = 36$): *S. chaucha* and *S. juzepczukii*. There is one pentaploid cultivated species ($5n = 60$): *S. curtilobum*.

There are two major subspecies of *Solanum tuberosum*: *andigena*, or Andean; and *tuberosum*, or Chilean. The Andean potato is adapted to the **short-day** conditions prevalent in the mountainous equatorial and tropical regions where it was originated. The Chilean potato, native to the **Chiloé Archipelago**, is adapted to the **long-day** conditions prevalent in the higher latitude region of Southern Chile.

17.4 AREA AND DISTRIBUTION

Potato growing zones are mainly distributed in **temperate** regions as well as **subtropical** regions, surprisingly in Northern hemisphere, although, it was originated in **Peru** of **Southern hemisphere**. It is grown extensively between latitudes of 45° to 57°N in temperate regions as summer crop, as well as 23° to 34°N, during winter—particularly in Europe and Asia. In subtropics and tropics, it is grown as successful crop in the hills up to 3500 m. About one-fourth of the global area is distributed in hilly regions above 1000 m above M.S.L. Maximum productivity is recorded in the temperate regions of China, Russia and Northern European countries.

China and India together produce 153 million tons (one-third of global production) out of 383 million tons produced at global level (2023). Till 1998, Russia, Germany and USA were leading in the production, indicating a shift in centre of the production from Europe to Asia.

In India, 60.14 million tons of potato is produced over 2.33 million hectare. The crop is distributed in Indo-Gangetic plains of Uttar Pradesh, Bihar, Gujarath, Madhya Pradesh and West Bengal–which together contributes 75% of the Indian area and contributes 80% of the production (Table 17.1). In this region, it is grown in **winter** between October to March. Around 5% of the area is distributed in Punjab, while hilly regions of the Northern India totally contribute around

5% of area and Southern peninsular region contribute around 6% of the total Indian area. In South India, potato is grown in high altitude plateau regions, where relatively cooler summer is found. In India, high productivity (up to 25–30 t/ha) is recorded in the Northern hilly/plateau regions, where its duration gets extended. In Southern plateau regions, the productivity is as low as 4–6 t/ha (Karnataka and Maharashtra).

Table 17.1 Area, Production and Productivity of Potato in Selected States (2021–22)

State	*Area (000 ha)*	*Production (million tons)*	*Productivity (t/ha)*	*State*	*Area (000 ha)*	*Production (million tons)*	*Productivity (t/ha)*
Punjab	110.0	2.85	25.90	West Bengal	447.45	12.40	27.72
Uttar Pradesh	622.50	16.16	25.96	Gujarat	127.80	3.70	28.95
Bihar	330.00	9.12	27.65	Haryana	29.54	0.78	26.49
Madhya Pradesh	158.40	3.58	22.65	Chhattisgarh	42.54	0.65	15.31
Jharkhand	49.28	0.69	14.18	Others	181.20	2.66	14.70
Assam	103.44	0.76	7.36	All India	2202.15	53.38	22.03

17.5 SOIL AND CLIMATIC REQUIREMENTS

- **Temperature:** It is a **cool season crop** requiring a temperature range of 25°C at germination. 20–24°C for growth and around 17–20°C for tuberisation coupled with less windy, low humidity and bright sunshine during growth. These conditions are satisfied during summer/spring in temperate and hilly regions, while it is grown as winter crop in plains and plateau regions of tropical and subtropical regions. If planted in low temperature (below 24°C), the sprouting is delayed. Similarly, if the crop experiences high temperature (25–30°C) or extremely low temperature (< 5°C) during tuberisation, the tuber growth is partially or fully stopped (possibly because high temperatures encourage higher respiration resulting in more degeneration of carbohydrate leading to poor accumulation of starch in tubers. Low temperatures will reduce the photosynthetic activity with low enzymatic activity).
- **Rainfall:** Potato requires well distributed light showers during the entire growth period, except at sprouting stage. But, heavy rainfall at any stage, including sprouting stage can promote more disease and pests. Initial stages of crop require lesser water, but post-tuberisation stages require more water. The crop will not withstand soil moisture depletion below 20–30%, thereby indicating more frequent irrigation for crop grown in the rainless period. The total water requirement ranges from 70–85 cm, depending upon the duration of crop.
- **Sunshine hours:** Since the flowering and reproductive growth are not leading to economic part of the crop, the length of day hardly matters directly. However, potato is a **short day plant** and flowers during short days. But, short days or low temperatures during early stages may prolong the duration. Longer days associated with higher temperatures during tuberisation may lead to poor development of tubers.

- **Crop duration:** The crop duration of potato is governed by climatic conditions. Cooler temperatures experienced in early stage will prolong pre-flowering period from 60–65 days to 110–120 days, as the photosynthetic rates are reduced. For this reason, in hilly regions (1000 –3000 m altitude) as well as temperate regions, the crop takes longer time to mature (as high as 130–150 days). But in plains, the duration as well as productivity is lesser (around 80 –100 days due to limited growing period).
- **Soil:** Although, potato crop can be grown in wide range of soils, the crop is more successful in well-drained, naturally loose, loamy and sandy loam soils, which are preferably rich in organic carbon status. Under no circumstances, the crop is successful in saline and alkali soils, but mild acidity (up to 5.5 pH) can be tolerated by the crop.

17.6 BOTANICAL DESCRIPTION OF PLANT

Potato is a perennial crop adapted to annual habits. The economic part **tuber** is a stem modification, but found under the soil. It is an extension of stolons, which grow horizontally from the base of the stem (Figure 17.1). The tubers have **eyes**, which are buds–capable to grow into independent plant. Various shapes and sizes of tubers could grow on potato plant–depending on the genotypic features.

The aerial part of the plant consists of angular pubescent or glabrous stems–greenish or purplish (Figure 17.2). Potato has branching habit which favours bearing more number of leaves and flowers. The leaves are alternate and compound grown along the stem in spiral arrangement. Nearly, 3–4 types of opposite leaflets are borne with single large terminal leaflet in every compounds leaf.

Figure 17.1 Angular stem with profuse green foliage.

Figure 17.2 Tubers from stolons.

Potato flowers are terminal clusters, with **pentapetalous**, **pentasepalous**; **bicarpellary** flowers having five stamen. They produce large percentage of sterile pollens and little fertile pollens. Fruits are generally not formed. However, the crop is cross-pollinated and if fertile pollen from other plants, the fruits (berries) may be produced. Potato roots are adventitious, arising from the base of sprout and roots are restricted to superficial layer of soil (20–25 cm). Rarely, some varieties may have roots to a depth of 80–90 cm.

17.7 VARIETIES

Highly diversified growing situations with varying ecological features require large number of varieties suitable for specific situations. Further, disease resistance and interaction of these diseases with varying climates have also called for different varieties for different states. In India, most potato varieties are released from Central Potato Research Station located in Kufri hill near Shimla. Hence, all of them carry **Kufri** as prefix. Important varieties recommended to different states are presented in Table 17.2.

Table 17.2 Potato Varieties Presently in Cultivation in Different Agro–ecological Zones of India

Potato Zones	*Duration*	*Recommended Varieties*
North-Western plains	Early	Kufri Chandramukhi, Kufri Jawahar, Kufri Khyati
	Medium	Kufri Pukhraj Kufri Anand, Kufri Arun, Kufri Badshah, Kufri Chipsona-1, Kufri Chipsona-2, Kufri Chipsona-3, Kufri Pushkar, Kufri Sadabahar, Kufri Sutlej, Khufri Surya, Kufri Bahar, Kufri Jyoti, Kufri Garima, Kufri Gaurav
West-Central plains	Early	Kufri Chandramukhi, Kufri Jawahar
	Medium	Kufri Arun, Kufri Anand, Kufri Badshah, Kufri Bahar, Kufri Chipsona-1, Kufri Chipsona-2, Kufri Chipsona-3, Kufri Frysona, Kufri Pukhraj, Kufri Pushkar, Kufri Sadabahar, Kufri Garima, Kufri Sutlej, Kufri Surya
	Late	Kufri Sindhuri
North-Eastern plains	Early	Kufri Arun, Kufri Chipsona-1,
	Medium	Kufri Chipsona-2, Kufri Chipsona-3, Kufri Lalima, Kufri Pukhraj, Kufri Pushkar, Kufri Sutlej, Kufri Surya, Kufri Jyoti, Kufri Kanchan, Kufri Bahar, Kufri Gaurav
	Late	Kufri Sindhuri
Plateau region	Early	Kufri Chandramukhi, Kufri Jawahar, Kufri Lauvkar
	Medium	Kufri Pukhraj, Kufri Badshah, Kufri Surya, Kufri Jyoti, Kufri Garima
North-Western hills	Medium	Kufri Giriraj, Kufri Himalini, Kufri Himsona, Kufri Girdhari, Kufri Jyoti, Kufri Shailja
North-Eastern hills	Medium	Kufri Giriraj, Kufri Himalini, Kufri Jyoti, Kufri Girdhari, KufrI Megha, Kufri Shailja
North Bengal and Sikkim hills	Medium	Kufri Jyoti, Kufri Kanchan
Southern hills	Medium	Kufri Giriraj, Kufri Himalini, Kufri Jyoti, Kufri Girdhari, Kufri Shailja, Kufri Swarna

Note:

Plains: Early (70–90 days), Medium (90–110 days) and Late (> 110 days)

Hills: Early (100–110 days), Medium (110–120 days) and Late (> 120 days)

Processing varieties: Kufri Chipsona-1, Kufri Chipsona-2, Kufri Chipsona-3, Kufri Himsona and Kufri Frysona

17.8 CROPPING SYSTEMS

Potato being short duration crop of plains, it could be grown in sequence with many crops under irrigated as well as rainfed situations. However, limitations of rainfall may restrict its inclusion under rainfed conditions. Common **sequential systems** involving potato are:

North Indian conditions under irrigation: All plain and plateau regions except hilly regions:

Maize–potato, rice–potato, soybean–potato, cowpea–potato–wheat, maize–potato–wheat, maize–potato–onion

South Indian rainfed hilly regions: It includes the following:

Potato–ragi, potato–sunflower–potato–rabi jowar

Potato could be grown as inter crop with autumn sugarcane. Similarly, many short duration vegetable crops like radish, carrot could also be grown along with potato.

17.9 NUTRIENT MANAGEMENT

Potato crop yielding 40 tons of tubers per hectare removes 175 kg N, 80 kg P_2O_5 and 310 kg of K_2O. The recommended doses of fertilisers vary from 75–200 kg N/ha, 75–129 kg P_2O_5/ha and 90–275 kg/ha K_2O/ha in different states–depending upon the nutrient status. In general, potato is **potassium loving crop**. It requires good fertile soil, hence, generally the crop is provided with copious amount of FYM (up to 30 t/ha).

Among the sources of nitrogenous fertilisers, combination of ammonical and nitrate forms like **Calcium Ammonium Nitrate (CAN)** is more suitable and it is followed by ammonical forms. Urea is less efficient when compared with other forms. Similarly, phosphorus in the form of Single super phosphate alone or in combination with rock phosphate in 1:1 ratio suits well for the potato. With respect to potassium, sulphate form is more suitable than chloride form as it has great influence on the quality of potato tubers.

Entire dose of phosphorus and potassium along with 50% of recommended nitrogen can be applied as basal dose and the remaining nitrogen can be applied at the time of earthing up. The fertilisers are applied in the furrows just below the seed tubers, either manually or mechanically with the help of tractor mounted fertiliser drill.

17.10 WATER MANAGEMENT

Depending on the season and extent/distribution of rainfall, potato crop is grown either as **rainfed crop** or **irrigated crop**. In Northern plains, it is mostly irrigated crop because it is grown in rain-free period, while in Southern plateau, Northern and Southern hilly regions, it is mostly grown as rainfed regions—as it is grown in rainy season. Wherever irrigated, it is popularly grown with ridges and furrow layout, with irrigations given frequently, not to allow the soil to have more than 20–30% depletion. Successful cultivation of potato has been taken up with drip irrigation or even with alternate row irrigation, preferably with fertigation. Over flooding of the crop may damage the tuber growth, but potato cannot withstand deficit situations.

On medium to heavy clay soils, 5–7 irrigations are sufficient, while in sandy/loamy soils, 8–20 irrigations may be necessary. The crop has water requirement in the range of 75–80 cm as it is considered to be sensitive to stress.

17.11 WEED MANAGEMENT

Weed management is tricky in potato. The initial 30 days are critical and in this stage, the weeds are controlled by intercultivation between rows at 15 and 30 DAS. But later operations may damage the tuber growth. In case, the weeds persist even after 30 DAS, hand weeding is preferred. If the weed infestation is severe or when the area is too large to control, the weeds by cultural operations, herbicides like Metribuzin @1 kg a.i./ha or Alachlor @4 l/ha could be used as pre-emergence spray.

17.12 SEED TUBER AVAILABILITY AND STORAGE

Tubers used as seed appear to be similar to potato meant for consumption. But, they are grown following separate technique called **seed plot technique**. Healthy seed tubers are most essential for successful commercial cultivation of potato. Such healthy seed tubers are grown by selecting tubers from a very reliable source with no history of disease. They are treated with 0.25% solution of Aretan.

Planting of seed crop is done between 10th and 20th October at a spacing of 60 cm × 20 cm in plains. A lower dose of N (80–100 kg/ha) is used to curb excessive vegetative growth. The crop should be examined at least thrice during crop period, and if any symptoms of disease are seen, such plants are uprooted and burnt to check the spread of disease. Normal plant protection measures are adopted. During mid-December, the irrigations are restricted and later withheld completely to avoid late season aphid attack. Tops are removed between 10th and 15th January. The tubers are left under the soil till January end and then harvested–so that the skin of tuber gets hardened. The tubers are hand harvested and carefully graded. Good tuber seeds are stored under **cold storage** for use in next season. As planting is done in *kharif* season in Southern India, they may be used within six months provided the tubers have no dormancy.

Dormancy in Potato
Seed tuber, exhibiting dormancy, cannot be used immediately, unless dormancy is broken. This can be done by the treatment with 1% Thiourea @10 lL solution for 50 kg tubers for one hour. The tubers are air dried in shade. Dormancy could also be broken by the treatment with 1 ppm gibberellic acid along with 3% ethylene chlorhydrin solution for one hour. Such treated tubers may have to be pre-conditioned and pre-sprouted in moist sand under polythene cover, at least for 72 hours to provide proper environment for sprouting.

17.13 METHODS OF PROPAGATION

Potato is propagated both by vegetative method as well as by seeds. More popular and widely practiced vegetative method involves the use of tubers as seed materials, as each tuber can sprout and result into few independent plants. The numbers of sprouts depend on the number of 'eyes' or 'buds' available on tuber. New shoot growth can occur as sprout grows from each bud. Tubers specially grown for seed purpose (called **seed tubers**), free of disease causing pathogens, are used for planting.

The flower of potato can be grown and produce fertile pollens under extended light period (by providing lighting facility in the green house) and such plants produce berry (fruit). The fruits are harvested and outer pulp is treated with HCl for few minutes and then washed to obtain the seeds, which are popularly called **true potato seeds (TPS)**. They are so called, because they are real botanical seeds out of fertilisation, while seed tubers are vegetatively propagated tubers (stem modifications). **True potato seeds can reduce the cost and drudgery of handling, transport and even planting of bulk quantity of seed tubers**, as 150 g seeds are sufficient for 1 hectare land. To use them for propagation, these seeds are sown—after fungicidal treatment—on a raised bed nursery (around 25–30 beds of 25' long, 4' wide and 4" high per hectare are required) at spacing of 6" × 4" and allowed to grow for a period of 60–70 days to develop a **tuber let**. These tuber lets are ultimately transplanted in the main field.

True Potato Seed Production
As the potato plants are likely to produce large quantity of sterile pollen, formation of berry is limited. To produce true potato seed, artificial facilities are needed to grow the pollen parent and female parent of suitable variety in separate adjacent plots under green house condition. The photoperiod is artificially extended by 4–5 hours and pollen production is ensured, so that berries of hybrid are able to be produced. For this, pollen is collected from male parents and dusted on the flowers of female lines by brush. The berries are harvested and pulp is softened by HCl treatment and washed in cold water and seeds are separated and dried under shade.

17.14 PLANTING METHOD AND SEASON

- **Seed selection:** It is very important to obtain seed tubers from a known reliable source, to achieve diseaseless crop. It is also advisable to sort the seed tubers to separate the disease affected tubers, before planting.
- **Preparation of tubers:** Seed tubers are invariably stored under cold storage and when such tubers are received, they have to be pre-cooled for 3– 4 days under room temperature before they are used as planting materials. The whole tubers of 30–50 g are invariably used for planting, when **early planting** is to be done (20–25 q whole tubers are necessary per hectare). Alternatively, cut tubers of 25 g each could also be used. They are cut in such a manner that each piece has 2–3 eyes. The tuber requirement may be 15–20 q/ha when pieces are used. The tubers are treated with 1% Chlorocin followed by rinsing. Then they are treated with 3% boric acid solution, before they are used for planting.
- **Spacing:** The potato tubers are usually planted on ridges separated by 60 cm at a distance of 15–20 cm.
- **Season of planting:** Early planting between 2nd week of September to 4th week of October is adopted in Western/Central/Eastern Indo–Gangetic plain.

Main planting: But, in North Western hill, planting of potato is done in the 1st week of April to 2nd week of May in the hill tops, while planted in 2nd week of December to 2nd week of January in lower hills. In North Eastern hills, summer planting is done during 1–2 week of March, while autumn planting is done during 4th week of August to 1st week of September. Main planting is done in October month in all the regions of Indo-Gangetic plains, while in plateau region of South India it is done in the 4th week of June to 1st week of July.

Late planting: In Indo-Gangetic plains, even late planting is done in 2nd or 3rd week of January in spring season or 2nd week of November to 4th week of December.

- **Planting methods:** Potato could be planted on ridges, which is very popular. Potato could also be planted on flat bed, by creating shallow furrows and ridges are also made after germination (when crop is 10 cm tall) and 2–3 earthing up are necessary to make the ridge thick. Such method is suitable for lighter soils.

17.15 PESTS AND DISEASES

- **Late blight:** Caused by *Phytophthora*, it can attack the crop any time after development of foliage. The symptoms include water soaked lesions towards margin of leave, which enlarge under ideal temperature and humidity. Later, lower side of leaves will develop cottony white growth, which turn black and the leaves start rotting. Similar symptoms may be seen on the stem and even on the tubers. Control measures include use of disease free tubers or spray of Ridomyl MZ @2 g/l and removal of infected material out of the field.
- **Early blight:** Caused by fungus *Alternaria*, early blight is more common than late blight at any time of the crop. Innumerous number of brown to black spots appears on the leaves with concentric rings, which later cause defoliation. Control measures include removal of affected plants or spray of solution consisting 2 g Mancozeb + 10 g urea per litre, when symptoms start.
- **Black scurf:** Caused by *Rhizoctinia*, this disease is common in hilly areas. The disease appears in two phases, viz., initial stem canker phase results in killing of sprouts and delay the emergence, and the later phase called **black scurf phase** can cause black crust on tubers and reduce the market acceptability. Use of treated tubers or treatment of soil with Brassicol @20–30 kg/ha or combination of both gives good results to control the disease.
- **Bacterial wilt:** Caused by *Pseudomonas*, this disease is common in mid-hills, and plateau regions. The symptoms include sudden wilting and drooping shoots associated with browning of vascular bundles in the stem and brown ring in the tuber. Even the eyes may turn black. Control measures include use of disease-free tubers from reliable source, early planting, blind earthing up or soil application of bleaching powder at the time of planting @12 kg/ha.
- **Wart:** Caused by fungus *Synchutrium*, the disease is characterised by the appearance of tumours/warts on tubers, stems or stolons. The use of disease free tubers or use of resistant varieties can reduce the incidence of the disease.

- **Leaf roll:** Most common disease in plains, the leaf roll is characterised by rolling of margins progressing towards midrib to involve entire lamina. The yield is drastically reduced. The control measures include avoiding small tubers for planting, destruction of affected plants or spray of Metasystox or Dimethoate @1 ml/l at 10–15 days interval.
- **Epilachna beetle:** It is a serious pest of potato. The beetles scrap chlorophyll and eat green parts from the leaves and leave only veins. Control measures include spray of Carbaryl @0.2% or dusting Carbaryl @30 kg/ha.
- **Cut worms:** The caterpillars cut the growing stem or twigs, leaves and cause heavy yield loss. They hide in the soil during night. Control measures include drenching the ridges with Chloropyriphos @2.5 ml/l using 300 lt/ha or application of Phorate granules @10 kg/ha or Carbofuran granules @30 kg/ha.
- **Aphids:** They are sucking insects, causing loss of turgidity in leaves, stem and tender shoots. The leaves of affected plants turn yellow and lose vitality. Besides this, the secretion from aphids attracts sooty mould and interferes in photosynthesis. Spray of Metasystox or Dimethoate @0.5 ml/l can control the pests.

17.16 CULTIVATION TECHNIQUES

- **Land preparation:** A well pulverised seed bed is necessary for potato. After the harvest of *kharif* crop, the land is ploughed with MB plough to a depth of 25 cm. It is followed by 2–3 harrowing criss cross and final levelling. Manuring @ 15–20 t/ha is useful before last operation is done.
- **Harvesting:** The harvesting of potato should be done after the **haulms start drying** and **yellowing**. At this stage, the haulms are cut and tubers are allowed to stay underground for 12–15 days. Later, the tubers are dug by **shallow diggers** like khurpi or by the use of tractor operated/bullock drawn diggers. After the tubers are collected, they left for 5–6 days under shade for thorough drying and then they are packed in special bags, which facilitate the aeration. A yield of 30 – 40 tons of tubers are expected from one hectare land. In hilly regions, the yields hardly cross 25 t/ha.

17.17 ANTI-NUTRITIONAL INGREDIENTS IN POTATO

Glycoalkaloids such as **solanine** and **chaconine** are found in dangerous proportion in the leaves, stem, sprouts and fruits of potato. Luckily, none of them are consumed by human beings. But, these compounds are likely to be **developed underneath the skin of tubers**—if they are exposed to sunlight during developments stage or when they are physically damaged or when they are over-aged tubers. These Glycoalkaloids may cause diarrhoea, cramps, and weakness and sometimes affect central nervous system.

General indication of presence of solanine is **greening of tubers** due to exposure to sunlight. A greened tuber may contain solanine to the tune of 12–20 mg/kg, while it may go up to 250–280 mg/kg, by excessive greening. The green skin alone may contain up to 2000 mg/kg solanine. Regular consumption of greened or even normal potato may lead to the intake of solanine without knowledge.

REFERENCES

American, J., Potato Research (2001), Vol. 78, No. 6, pp. 403–412.

Anonymous (2012), International Potato Center: World Potato Atlas (India) https://research.cip.cgiar.org/confluence/display/wpa/India: International potato centre-India.

Anonymous (2012), Potato, http://www.krishisewa.com/cms/articles/production-technology/316-potato-production.html.

Anonymous (2013), Potato, http://www.ikisan.com/Crop%20Specific/Eng/links/ap_potatosoiland climate.shtml. Potato Production Technology for Plateau and Southern Hills.

Anonymous (2013), Potato Varieties Developed, CPRI, Shimla, http://cpri.ernet.in/?q=node/91. Anonymous (2013), Potato, Wikipedia, http://en.wikipedia.org/wiki/Potato.

Anonymous, 2022, Agricultural statistics at a glance, DAFW, GoI, pp. 29.

FAOSTAT, 2023, https://www.fao.org/faostat/en/#data/QCL.

Pandey, R.P. (2002), *The Potato*, Kalyani, Ludhiana, pp. 9–147.

Chapter 18

Tobacco (*Nicotiana tabacum* L.)

18.1 ECONOMIC IMPORTANCE

Tobacco is world's **controversial commercial crop**, often economic benefits of it outweigh the health hazards. Its cultivation and use is a controversial subject for many countries, but its international trade has increased over the years–providing fillip to unabated cultivation. The tobacco leaves are important economic product used in diversified ways of intoxicating the human nervous system. It is considered as **narcotic crop** along with completely banned opium crop, which is often illegally grown in many countries.

Tobacco leaves are invariably used after specific **curing process**—each curing process is unique for specific range of products desired to be used for specific set of customers. Thus, the tobacco leaves end up in many products like cigarette, cigars, bidi, cheroot, hookah, chewing tobacco and snuff. Besides leaves, tobacco seeds are used as a source for nicotine sulphate–common ingredient of many skin ointments. In the past, tobacco was traditionally used for medicinal purposes and even religious purposes by aboriginal Native Americans. Although controversial, tobacco is such an important commercial crop, farmers, tobaccos traders, tobacco products manufacturers and even government have nurtured its production and sale for number of benefits–throughout last five decades amidst all controversies. However, use of tobacco has definite adverse impact on the health of human beings—as it can cause lung cancer, respiratory disorders, liver/stomach dysfunction, mouth cancer and many more miseries to human life. Essentially, an alkaloid called **nicotine** contained in it is an intoxicant and stimulates central nervous system and increase the blood circulation temporarily providing the relief to the person consuming tobacco.

Tobacco has been a great revenue earner for the Government, as it attracts excise duty besides commercial tax–when it is traded and transformed before it reaches the consumers. It is also a foreign exchange earner, as substantial crop of tobacco is exported to different countries. Whenever, the health issues arise, there is pressure on the Government to ban the tobacco, which is invariably negated by most Governments—including Indian.

What Causes Cancer, when Tobacco is Smoked?
Contrary to common understanding, it is not alkaloid **nicotine** that causes cancer, but tar generated during smoking that causes cancer. The smoke consists of carbon dioxide and carbon monoxide which develops tar—that is deposited in the lungs, impairing the functioning of lungs, which later leads to many physiological disorders leading to cancer.

18.2 ORIGIN AND HISTORY

Tobacco is believed to be originated in **Ecuadorean** and **Peruvian Andes mountain range**, where it had been growing for at least five thousand years, although, tobacco plant was known well before that. Later it spread to the Eastern part of Mexico and many countries in Latin America. But, the world did not know about tobacco till **Columbus reached North America** in the 15th century. He carried tobacco to Europe. Although, native Americans practiced smoking tobacco, Europeans popularised smoking tobacco, as they could standardise the curing process. In realms of history, tobacco is associated with many fascinating facts–including tobacco-based currency exchange, tobacco-based trading and manufacturing, tobacco standards (like gold standards for monetary transactions). Tobacco is also associated with monasteries, battles, industrial uses, medicinal uses and host of other historical issues of civilisation.

History of Smoking
Tobacco was not popularly used for smoking till sixteenth century, but mainly used for medicinal and religious purposes. Occasionally, it was smoked in pipes, to carry one's thoughts to Heaven, as native Americans believed that tobacco was the gift of God. After introduction to Europe, the whole world learnt SMOKING, now considered as universal, most widespread vice.

18.3 DISTRIBUTION

Tobacco is grown in more than 95 countries, over 3.2 million hectares in the world, with 6.0 million tons of unmanufactured tobacco. Its area is mainly distributed in the following countries–Tanzania (1.6 lakh hectare), USA (0.75 lakh hectare), Malawi (0.78 hectare), Indonesia (1.9 lakh hectare), India (4.2 lakh hectare), China (1.0 million hectare), Turkey (1.08 lakh hectare) and Brazil (3.2 lakh hectare). Notable other countries known for tobacco cultivation are Argentina, Bangladesh, DPR Korea, Lebanon, Mozambique, Pakistan, Philippines and Thailand. (FAOSTAT, 2023). India is the second largest producer and exporter of tobacco in the world.

In India, tobacco is grown in specific pockets of many states and rarely spread throughout any state except Andhra Pradesh.

18.4 BOTANICAL DESCRIPTION OF PLANT

The genus *Nicotiana* has more than 70 species, out of which two species *tabacum* and *rustica* are most popular, due to their intoxicating properties. It belongs to family *Solanaceae*. It is an **annual shrub**, growing to a height of 1–1.5 m, with erect hollow green stem and supporting branches which generally grow to give conical shape to the plant. The leaf development on main stem as well as the branches is in acropetal fashion. The leaves are broad velvetty for touch, glandular, petiolate, caudaceous in shape with prominent midrib (Figure 18.1). The leaf may attain 25–50 cm long and 20–35 cm wide. The inflorescence is spike borne terminally with more than 150 tiny flowers (Figure 18.2). The petals of *N. tabacum* are pinkish, where as those of *N. rustica* are greenish white or yellow. Tobacco is a **self-fertilised crop**, but cross-pollination may occur to the tune of 5–10% by insects. The seeds of tobacco are extremely small. A single fruit may contain more than 8000 seeds.

Figure 18.1 Inflorescence of tobacco.

Figure 18.2 Leaf growth in tobacco.

Differences between *N. tabacum* and *N. rustica*
***N. tabacum*:** Plants are taller (1.5–2 m), leaves are longer and narrower, sessile or petiolate; flowers are reddish, pinkish or white; essentially used for smoking or chewing purposes. Used to manufacture cigarette, cigars, chiroots, bidi, chewing and snuff.
***N. rustica*:** Plants are bushy and shorter (0.9–1.2 m); leaves are larger and broader and ovate in shape, always petiolate. Flowers are in clusters and are dull greenish—yellow. Used for the manufacture of hookah, chewing and snuff purposes.

18.5 SOIL AND CLIMATIC REQUIREMENTS AND SPREAD IN INDIA

Tobacco crop is **extremely sensitive** to both soil and climate, not only in respect to yield, but with respect to different curing qualities as well as various consumers' preferred qualities. Soil and climatic requirements of tobacco crop cannot be generalised like other crops, because different types of tobacco are grown in different types of soil and climate. The quality of tobacco leaves grown in specific agro-climate carries specific quality—desirable for specific tobacco product, having specific client group. In general, it can be grown in *kharif* or *rabi* season on wide range of soils, preferably **well drained** and **slightly acidic in reaction** (a pH of 5.5–6.5). It grows well under **tropical** conditions and requires **warm sunny days**. Humid climate accentuates diseases. The soil requirement for different categories of tobacco are briefly summarised as:

- **Flue Cured Virginia (FCV) tobacco for cigarettes:** It is principally grown in four different situations in India, namely:

  1. **Traditional black soils** like clay loams, silty clay loams and clays, highly clayey (50–80% clay) throughout the profile, slightly **alkaline in reaction** (pH 7.5 to 8.8), calcareous, highly fertile, having high water holding capacity with very poor drainage. Tobacco is grown on **stored soil moisture** as a post-monsoon crop during winter. The tobacco growing area is spread to the East and West Godavari districts, Khammam, Krishna and Guntur districts and Prakasam and Nellore districts of Andhra Pradesh.

2. **Northern light soils** are spread to the East Godavari, West Godavari and Khammam districts of Andhra Pradesh. They are **sandy loams** to **loamy sands**, **slightly acidic** with very low exchangeable cations, **low water holding capacity**, and **poor fertility** status with very good drainage. Tobacco is grown in these soils under irrigated conditions during winter.
3. **Southern light soils** spread in Prakasam and Nellore districts of Andhra Pradesh and are **red loamy soils**, **neutral in reaction**, low to medium fertility status, **moderately well drained**, moderately low permeability, with **moderate water holding capacity** and low to medium cation exchange capacity with more than 75% base saturation. Tobacco is grown during winter on stored soil moisture from North East monsoon rains.
4. **Karnataka light soils** are **red soils**, which may have yellow to deep red in colour, **loamy sands** and **sandy clay** in texture (clay: 10 to 25%), **low in inherent fertility**, **slightly acidic** in reaction with fairly good water holding capacity. The soils are well-drained and highly leached. Here, tobacco is grown during kharif as rainfed corp.

- **Burley tobacco:** It is grown in the East and West Godavari districts of Andhra Pradesh and the soils are **sandy loams** (surface) and **loams** (sub-soil) and the pH is **neutral** to **acidic** (5.5 to 6.5 pH). The soils are very low in soluble salts, chlorides and light in texture. The organic matter content, available nitrogen and available phosphorus status is low.
- **Natu tobacco:** The crop is raised on **medium** to **heavy black soils** with a pH range of 7.0–8.5. These soils are poor in available nitrogen, organic carbon, medium to high in phosphorus and high in available potash. It is also grown on **sandy** to **sandy loams** with low organic carbon, low in available N, P and K as an irrigated crop.
- **Lanka tobacco:** Lanka tobacco is exclusively grown on the river banks and deltaic islands of Krishna and Godavari rivers in Andhra Pradesh. The soils are **alluvial** and **highly fertile**. Lanka tobacco is cultivated year after year on low lying lankas (islands). Lanka tobacco raised on high level lands on the other hand, requires two or three irrigations.
- **Bidi tobacco:** In Gujarat, bidi tobacco is grown in Anand, Petlad and Nadiad taluks and in some parts of Vadodara District on soils sandy to sandy loam soils called **Goradu**. The banks of rivers Mahi and Sabarmathi are also considered to be the best soils for bidi tobacco. It is mostly grown under **irrigated conditions**. In Karnataka, bidi tobacco is mainly grown in Nipani on the banks of river Krishna and its tributaries, on **black silt loams soils** which have good water retentive capacity. In Maharashtra, cultivation of bidi tobacco is concentrated in the fertile, **medium to deep** and **black soils** in upper Krishna basin in Kolhapur and Sangli Districts. In Karnataka and Maharashtra, bidi tobacco is a **rainfed crop**, though in Nipani area it is grown under irrigation from wells. Bidi tobacco is also grown on small scale in Koraput District of Odisha and in Kurnool District of Andhra Pradesh.
- **Chewing and hookah tobacco:** Chewing and hookah types of tobacco are grown on different types of soils, ranging from **sandy loam upland soils** of Vaishali, Samastipur, and Muzaffarpur to **medium** and **heavy paddy growing soils** of Purnea, Katihar and

Saharsa. But the main tobacco growing area in North Bihar consists of nonsaline calcareous sandy loam soils of alluvial type, with soil pH ranging from 7.5–8.5, with $CaCO_3$ as high as 45%.

Most of the tobacco growing soils are low in organic matter, available N and P, medium in K content and mostly alkaline in reaction (pH 8.2–8.4). The **illite** and **muscovite** are the dominant minerals responsible for the enrichment of such soils. Available phosphorus content in these soils is low. Well-drained, neutral to slightly alkaline soil with good nutrient supplying capacity is considered ideal for chewing and hookah tobacco cultivation.

The chewing tobacco cultivation in Tamil Nadu is concentrated in the districts of Coimbatore, Salem, Thanjavur, Madurai, Tirunelveli and South Arcot. Chewing tobacco in Tamil Nadu is grown in a wide range of soils. But in any location, soils of comparatively alkaline in nature and irrigated with waters having high salt content are assigned to chewing tobacco.

- **Wrapper, Filler, Jati and Motihari tobacco:** Motihari (*N. rustica*) and jati, wrapper and filler (*N. tabacum*) tobacco types are cultivated in Cooch Bihar, Jalpaiguri, Mushidabad and Malda districts of North Bengal region. The soils of the area are alluvial flood plains in origin formed from the material deposited by the Ganga Brahmaputra river waters. The soils are **sandy loams** and **silt loams**, light in texture, whitish grey to grayish in colour, well-drained and well-aerated. The water and nutrient holding capacity of the soils is poor because of their coarse texture. Soils of North Bengal are of recent origin, and are less weathered. The alluvial sandy loams of North Bengal are acidic in soil reaction (pH 5.1–6.4), high in available P status and low to medium in available K.

Tobacco is successfully grown under **tropical climate** in **any season**. It is also grown in sub-tropical and temperate climates in spring by avoiding the cold temperatures. Normally, it requires about 100–120 days, frost-free climate with an average temperature of 80°F, to mature. In India, tobacco is grown under a very wide range of conditions from the coastline to an altitude of 3,000 feet. In the South, the crop is raised in winter from October to March when the temperature is moderate, but in Punjab, it is grown as an early summer crop. In the Eastern and Western parts of the country, it is grown between September and January.

- Ideal conditions required for the successful production of high quality leaf are: A liberal and well-distributed rainfall during active vegetative growth stage. Long day lengths.
- A mean temperature of 26°C during growing season.
- A high relative humidity of 70–80% during early period.

18.6 DIFFERENT TYPES OF TOBACCO IN DIFFERENT STATES

Ten distinct types of tobacco are grown in 15 states of the country. They are FCV (Flue Cured Virginia tobacco), Burley tobacco and Oriental tobacco used for cigarette manufacturing as well as non-cigarette tobaccos like bidi, chewing, hookah, natu, cheroot and cigar types.

- **Andhra Pradesh:** It grows Flue cured cigarette tobacco, Burley tobacco, Natu tobacco, Lanka tobacco and Bidi tobacco

- **Gujarat:** Bidi tobacco
- **Karnataka:** Flue cured cigarette tobacco, Bidi tobacco
- **Bihar:** Chewing and hookah tobacco
- **Maharashtra:** Bidi tobacco
- **Odisha:** Bidi tobacco
- **Tamil Nadu:** Chewing tobacco
- **West Bengal:** Wrapper, Filler and Motihari tobacco

Facts about Bidi in India
• The word **bidi** is surprisingly a modification of the word **beeda**, areca nut wrapped in betel leaves. Beeda has no ingredient of tobacco.
• Smoking bidi accounts for 50% of the total tobacco consumption in India.
• Bidi have more nicotine and it releases more carbon monoxide and tar than cigarette, hence is more risky to cause oral, and lung cancer.
• Bidi industry supports millions of families and brings huge revenues in central excise, but kills crores of people every year.

Explanation about different types of tobacco and tobacco products:

- **Bidi:** It is thin Indian desi cigarette filled with tobacco leaf bits wrapped in tendu leaves.
- **Cigarette:** It is finely packed tobacco leaf powder rolled in paper with or without butt.
- **Gutka:** Chewing tobacco is mixed with areca nut powder, slaked lime, paraffin and cashew and presented in small packets. It is consumed in small quantities regularly.
- **Snuff:** Nicely powdered tobacco leaves used to inhale through nostrils.
- **Hookah:** It is a single or multi-stemmed instrument for vaporising and smoking flavoured tobacco called **shisha**, in which the vapour or smoke is passed through a water basin, often glass-based, before inhalation.
- **Chewing tobacco:** The tobacco leaves are powdered and chewed with areca nut.
- **Burley tobacco:** Light air cured tobacco used in cigarette manufacturing.
- **Flue cured tobacco (also called Virginia Flue cured tobacco):** It is cured under controlled temperatures using hot air in enclosures. Invariably, VFC tobacco is used to make cigarettes.
- **Lanka tobacco:** Air cured tobacco is used for cheaper cigarettes.

Note: All forms of tobacco are dangerous to human health and can cause different types of cancers.

18.7 CURING OF TOBACCO

Tobacco leaves are not used for any of the tobacco products, unless they are cured. Curing imparts desirable properties in them, by changing their colour, moisture, texture, brittleness and many biochemical ingredients—which decide host of other properties influencing their aroma, flavour, taste, burning properties, etc. Hence, curing of tobacco is an inseparable part of tobacco cultivation to convert the raw tobacco leaves into useful commercial products.

In brief, curing involves treatment of tobacco leaves by various physical processes like smoking, heating, drying, all of which essentially remove the moisture from the leaves under a specific set of conditions. Different curing methods achieve different qualities in final product and therefore, **different products require different methods of curing**. Out of all tobacco curing methods, Flue curing method has special significance and involves more extensive procedures.

Tobacco curing is also known as **colour curing** because when tobacco leaves are cured with, the intention is to change their colour and reduce their chlorophyll content.

Different types of curing tobacco leaves are:

- **Flue curing:** It is invariably used to produce cigarette tobacco. It is achieved in specialised structure called **barn**. It is specially constructed structure usually to a height of 4–5 m and dimensions of 5 × 5 m or 4 × 4 m or 8 × 5 m, along with provision to generate hot air by having a source to burn wood/charcoal/husks/residues. Hot air is passed in the entire barn by a network of pipes. Facility is made to open and close the two layers of windows to regulate the temperature in the barn. The barn is filled with harvested leaves (called **charging**). These leaves are tied to bamboo poles of 1–1.5 m and these poles are stalked horizontally by proper support structures in the entire barn in such a manner that young/greener leaves are charged in the top and mature/yellowish/pale coloured leaves are charged in the bottom.

Flue curing involves four stages, each stage calling for specific regulation of temperatures, which is achieved by regulating the heat source and movement of air. The four stages are:

1. ***Yellowing stage*:** This extends from 36–48 hours. In this stage, the leaves are cured by air at a temperature of 32°C and humidity of 71%, and the temperatures are risen by 3°C—every 6 hours to finally achieve 48°C. During this stage, both top and bottom windows are kept open slightly in the beginning and is gradually increased till final temperature is achieved. The leaves turn to yellow colour during this stage.
2. ***Colour fixation stage*:** To fix the yellow colour, the temperature is gradually raised by 3°C every 2–5 hours till temperatures reach 51°C. As the temperatures rise, sufficient aeration should be provided by opening the windows gradually and at 51°C, all the windows are opened. The humidity of air will be around 30–55%. It takes about 16–24 hours to fix the colour.
3. ***Leaf drying stage*:** Leaves can be dried by increasing the temperature from 51°C to 60°C, by gradually increasing the temperature. The leaves are likely to develop brown spots, if the temperatures are not maintained in the desirable range. It requires 25–30 hours to complete this stage.
4. ***Midrib drying stage*:** The temperature is further raised from 60°C to 71°C gradually by rising 3°C every 2–3 hours and keeping bottom windows closed and top windows open. After the temperatures reach 71°C, all the windows are opened. This stage will take 10–12 hours.

After completion of the curing process, all the windows of the barn are opened and whole barn is cooled down to normal temperatures in 1–2 days. About, 7.5 tons of fire wood or 4.5 tons of coal are needed to cure the leaves from one hectare land. The leaves are allowed to absorb atmospheric humidity for 2–3 days and achieve softness. At this stage, the leaves are removed from sticks and transferred to the grading yard.

- **Bulking:** It refers to **collecting** the cured leaves and **stacking** them with frequent change over to ensure that they do not develop fungal attack or do not develop any heat. Invariably, the cured tobacco leaves are bulked for a period of 8–10 days and then graded.
- **Grading:** The cured tobacco leaves are graded according to the colour, grains developed on lamina, thickness, plasticity, moisture, size and shape of leaves, etc. into as many as 60 grades. Usually, at farmers' level, only few grades are identified. The cured leaves are baled according to the grades and kept ready for marketing.
- **Air curing:** In air-curing, entire tobacco plant is hung upside down in well-ventilated barns (Figure 18.3) and allowed to dry over a period of four to eight weeks. Air cured tobacco is low in sugar, which gives the tobacco smoke a light, sweet flavour, and high nicotine content. **Cigar**, **hookah**, **wrapper** and *burley* **tobaccos** are air cured.
- **Sun curing:** Sun curing involves drying of tobacco leaves under bright sunlight for 15–20 days, by tying the pre-wilted leaves (they are wilted in field for 1–2 days) on a bamboo pole (Figure 18.4). In India, sun curing is used to produce so-called **white snuffs**, which are fine, dry, and unusually potent. This method is used in Turkey, Greece, Bulgaria, Macedonia, Romania and Mediterranean countries to produce oriental tobacco. Sun cured tobacco is low in sugar and nicotine and is used in cigarettes. Usually, **natu**, **chewing**, **hookah** and **bidi** tobacco are sun cured.

Figure 18.3 Air cured tobacco leaves.

Figure 18.4 Sun cured tobacco leaves.

- **Smoke curing:** (also called **fire curing**): The tobacco plants are hung upside down in large barns where fires of hardwoods are kept on continuous or intermittent low fire—which produces hot smoke, which is passed over stacked tobacco plants for a period of three days and ten weeks, depending on the process and the tobacco quality desired. Fire curing produces a tobacco, low in sugar and high in nicotine. Pipe tobacco, chewing tobacco, and snuff are fire cured.

- **Pit curing:** This is invariably followed in India exclusively for **chewing tobacco.** Plants are wilted overnight and loaded in a pit of 3 m diameter and 3 m depth plastered with cement. The plants are laid in uniform layers and trampled gently to allow **anaerobic fermentation** by sealing with gunny cloth and mud plaster from the top. The anaerobic fermentation leads to a typical smell and dark colour to the leaves. After, 24 hours, they are transferred to other pit where they are kept for 48 hours to complete the curing process. Sometimes, small quantity of common salt and water are added to impart the specific flavour.

Cured Leaf Yield
Unlike other crops, the productivity of tobacco is expressed as **Cured Leaf Yield (CLY)** and not in terms of freshly harvested leaf yields, as cured leaves are marketable products. Generally, the ratio of fresh and cured leaf yield is 1:0.3–0.4.

18.8 VARIETIES

Large number of tobacco varieties are released by the Central Tobacco Research Centre, Rajahmundry, Andhra Pradesh, and its sub centers at different places as well as many state agricultural universities in the different states. These varieties are released for specific applications and for specific regions. Some of the important varieties are listed in Table 18.1.

Table 18.1 Varieties of Tobacco Grown for Different Uses in Different Area

Name of the Variety	*Used as*	*Yield Potentiality CLY/ha*	*Areas of Adoption*
CTRI Spl. (MR)	FCV tobacco	1200	Black soils and SLS of Andhra Pradesh
Godavari spl.	FCV tobacco	1520	Black soils and SLS of Andhra Pradesh
Swarna	FCV tobacco	1450	KLS of Karnataka
Jayashree	FCV tobacco	1550	Black soils and SLS of Andhra Pradesh
Bhavya	FCV tobacco	2000	KLS of Karnataka
Hema	FCV tobacco	1560	Black soils of Andhra Pradesh
Gouthami	FCV tobacco	2000	Black soils and SLS of Andhra Pradesh
Ratna	FCV tobacco	2000	KLS of Karnataka
Hemadri	FCV tobacco	2500	Black soils of Andhra Pradesh
Siri	FCV tobacco	2900	Black soils of Andhra Pradesh
KST 28	FCV tobacco	2000	KLS of Karnataka
CH 1	FCV tobacco	2900	Northern Light Soils
CH 3	FCV tobacco	2700	KLS of Karnataka
N 98	FCV tobacco	2200	SLS of Andhra Pradesh
FCH 222	FCV tobacco	3000	KLS of Karnataka
GT 7	Bidi tobacco	2500	Rainfed bidi tobacco regions of Gujarat

Name of the Variety	*Used as*	*Yield Potentiality CLY/ha*	*Areas of Adoption*
GTH 1	Bidi tobacco	3650	Irrigated bidi tobacco region of Gujarat
GT 9	Bidi tobacco	3000	For Gujarat area
Bhavyashree	Bidi tobacco	2000	Irrigated regions of Karnataka
MRGTH 1	Bidi tobacco	3200	All bidi tobacco regions
ABT 10	Bidi tobacco	3500	For Karnataka bidi tobacco regions
Vedaganga 1	Bidi tobacco	3400	For Karnataka bidi tobacco regions
Abirami	Chewing type	3700	For Tamil Nadu chewing tobacco area
Kamatchi	Chewing type	4000	For chewing tobacco regions of Tamil Nadu
Sangmi	Chewing type	3200	Erode and Salem District of Tamil Nadu
Meenakshi	Chewing type	3500	Interior regions of Tamil Nadu
Kaviri	Chewing type	2500	For coastal regions of Tamil Nadu
Manasi	Chewing type	1700	North Bengal region
Lichchavi	Chewing type	3000	Northern Bihar
Vaishali sp.	Chewing type	2700	Bihar
Margadham	Chewing type	3000	Smoke cured tobacco area of Tamil Nadu
Dharla	Hookah and chewing	2700	Suitable for North Bengal region
GCT 3	Hookah and chewing	5900	For North Gujarat
Tungabhadra	Oriental	700	For rainfed regions of Andhra Pradesh
Torsa	Motihari	2200	For Cooch Behar tract
Bhairavi	Natu tobacco	2600	For cigarette natu tobacco producing areas
Natu special	Natu tobacco	1600	
Sendarpatti spl.	Cheroot tobacco	2100	For Salem region
Lanka special	Cheroot tobacco	2780	For East Godavari District of Andhra Pradesh
Bhavani spl.	Cheroot tobacco	2850	For Coimbatore District
Krishna	Cigar wrapper	2250	For North Bengal region
S 5	Cigar wrapper	1500	
Banket A 1	Burley tobacco	1800	For light soils of East Godavari and Vishakhapatnam Districts of Andhra Pradesh
Burley 21	Burley tobacco	1700	

18.9 GENERAL CULTIVATION PRACTICES

- **Land preparation:** The lands will have to be prepared well by 1–2 ploughing with MB ploughs followed by 2–3 harrowing and final levelling. Application of adequate manures is important (8–10 t/ha) for all rainfed tobacco as well as any tobacco for bidi, chewing, hookah and snuff types.
- **Layout:** Invariably, tobacco crop is established on ridges and furrow layout.
- **Preparation of nursery:** Tobacco crop will have to be established by establishing nursery, irrespective of types of tobacco. Following steps are necessary to be followed for nursery:
  - Well-drained and deeply cultivated fertile soils are selected and raised bed of 1 mt width at convenient length (3–5 m) and 15 cm height are prepared after the application of manure @15–20 kg/bed. Before manure application, it is advisable to spread the paddy straw on the bed thinly and burn the entire bed to get rid of pathogens and insects. This process is called **rabbing.**
  - It is advisable to change the location of the nursery every time to avoid soil borne diseases and root knot.
  - For every bed, 205 g of CAN./Ammonium sulphate + 500 g Super phosphate and 250 g of SOP are added and thoroughly mixed. At the rate of 25–30 g/ha of transplanted fields, seeds of reputed sources are mixed with sand and spread on the beds at least 6–8 weeks before the planting season. Cover the beds with paddy or other straw.
  - After the germination (5–8 days), the straw cover will have to be removed for better growth of seedlings. If the seedlings are too crowded, it may be useful to thin them.
  - Irrigate the beds lightly every alternate day for 30 days and later every day.
  - If the soil is deficient in zinc, it is useful to apply $ZnSO_4$ at nursery state itself @2 kg for every bed.
- **Planting:** Planting of seedlings will have to be done when soil moisture is ideal, but no continuous rains are expected. Ideally, the planting has to be done at the top of the ridges @ one seedling at each spot. In rainfed situations, if the soil moisture is insufficient when planting is being done, transplantation may be done by spot irrigation.

To avoid the gaps due to the death of seedlings after transplantation, raising the seedlings in polyethylene bags is advisable, so that gaps could be transplanted by such seedlings.

- **Fertiliser application:** Both basal and top dressing should be done by spot application 5 cm away from the spot of transplantation.
- **Inter cultivation and weed control:** Since the tobacco seedlings are transplanted with sufficient space, it is always advisable to control the weeds by intercultivation. In regions where the weed intensity is very high, herbicides like Pendimethalin or Alachlor @1–1.5 kg a.i./ha as pre-planting spray. Orobanche cernua (Broom rape) is a holoparasite on tobacco (Figure 18.5), causing serious damages, in endemic areas. It can be controlled by the crop rotation with non-solanaceous crops, or trap cropping with sorghum/cow pea, sunn hemp/greengram, induced suicidal germination by chemicals

like Nijmegen 1, chemical control by applying maleic hydrazide or fungal herbicide Fusarium oxysporum.

Figure 18.5 Orobanche (broom rape) attacking tobacco.

- **Topping and desuckering:** Depending on the type of tobacco, the apical shoot of the plant is cut at specific stage to promote the growth of leaves (in terms of area, thickness and texture, etc). Topping invariably results in the growth of shoots called **suckers** in leaf axils. They also have to be removed by hand plucking them or by the application of chemicals like Decanol. Removal of suckers is called **desuckering**.

18.10 PESTS AND DISEASES

- **Damping off:** A nursery disease. Young seedlings decay or disappear due to **rotting**. Older seedlings become shrivelled at collar. To control it, provide good drainage or rabbing the beds with husk for a depth of 15 cm or spray 0.2% Ridomyl.
- **Black shank:** Nursery or main field disease. Black lesions on stem starting from ground may lead to drooping and fall of leaves. Plants may die. Control measures include rabbing (paddy husks to a height of 15–20 cm are placed on beds 8–10 days before sowing and burnt). Soil drenching with 0.2% Blitox may also be useful.
- **Frog eye spot:** It may appear late in nursery or in main field. The symptoms include oval, brown spots with ashy centre (dead tissues) appear on the leaves. Control measures include spraying 0.2% Zineb frequently in the nursery or the spray of Bavistin @0.03% after transplanting.
- **Powdery mildew:** After transplantation, the leaves appear dusted with white powder in patches, which may spread. It affects curing qualities and reduces marketability. It can be controlled by dusting with sulphur @40 kg/ha or spray 0.02% Carbendizim.

- **Anthracnose:** It appears in nursery, where small irregular whitish spots develop on the leaves, which coalesce to form bigger spots. The leaves dry. Or, brown lesions girdle the whole stem. Control measures include rabbing the beds or spray 0.03% Carbendizim.
- **Tobacco mosaic:** Most widely spread viral disease occurring in nursery or main field. Leaves will develop mosaic pattern of light green and dark green areas. Young developing leaves may get mottling, blistering, and pluckering. Affected plants may be stunted and pale. The disease will **reduce the market values of the leaves**. Control measures include spray tannin acid @1% at nursery stage, and the removal of affected seedlings as early as possible.
- **Root knot:** It attacks in the nursery and the main field. Attacked plants develop knots or gall on the roots. The leaves turn yellow and show poor growth. Plants may wilt and die. Control measures include deep ploughing in summer, rabbing the beds, keeping the fields free of weeds, crop rotation with groundnut, cotton, sesamum, chillies or application of phorate @15 kg/ha.
- **Tobacco caterpillar:** Caterpillar cuts the seedlings and feed on the young leaves and finally all the seedlings may get defoliated. Spray of Monocrotophos @1 ml/l at least two to three times is useful to control the pest.
- **Stem borer:** It feeds on the tissues inside the stem and midribs. Swelling appears on the attacked stem or midrib resulting into stunting and unusual branching of the seedlings. It may spread to the main field through the seedlings. Spray of Monocrotophos @1 ml/l at least two to three times is useful to control the pest.
- **Ground beetles:** They cut freshly transplanted seedlings causing the gaps. Spray of Chlorpyriphos @1.8 ml/l at the time of transplantation or application of Pongamia cake powder or neem powder cake @5 g. with handful of sand at the base, immediately after transplanting can bring control of the pest.
- **Green peach aphid:** The swarming population suck the sap from leaves turning the leaves sicky and development of sooty mold making the leaves unfit for curing. Control measures include spray Oxydemeton methyl @1.25 ml/l.

18.11 SPECIFIC CULTIVATION PRACTICES FOR DIFFERENT TYPES OF TOBACCO

- **Flue Cured Tobacco:** It includes the following cultivation practices:
    - ***For black soils*:** Spacing of 80 × 60 cm for transplantation; Planting at October middle to November first week; FYM @7.5 t/ha; Fertiliser @20–30:50:30–50 kg N, P_2O_5 and K_2O/ha; Topping at flower head; harvesting by priming.
    - ***For northern and southern light soils*:** Spacing of 100 × 60 cm for transplantation; Planting in October; 6–12 t/ha FYM; Fertilisers @40:60:60 Kg N, P_2O_5 and K_2O/ha (N and K in equal splits after 40 and 60 DAP); topping at flower head; harvesting by priming
    - ***For Karnataka light soils*:** Spacing of 100 × 60 cm for transplantation; Planting at May end or first week of June; 5–6 tons FYM/ha; Fertilisers @40:80:80 kg N, P_2O_5

and K_2O/ha (N and K in equal splits after 40 and 60 DAP); topping at flower head; harvesting by priming.

- **Burley Tobacco:** The following cultivation practices are used:
    - ***For light soils*:** Spacing of 100 × 50 cm for transplantation; Planting at July end to August first week; FYM @10 t/ha; Fertilisers @60:40:40 kg N, P_2O_5 and K_2O/ha (N in 3 equal splits after 40, 60 and 80 DAP and K in two equal splits at 40 and 60 DAP); Topping at 12–24 leaves; harvesting by priming.
    - ***For black soils*:** Spacing of 90 × 90 cm for transplantation; Planting at October end; 15 tons FYM/ha; Fertiliser @40: 15–50: 10–25 kg N, P_2O_5 and K_2O/ha; topping at 14–16 leaves; harvesting by priming.
- **Natu Tobacco:** Spacing of 90 × 90 cm for transplantation; Planting in October middle to November first week; Use of groundnut/linseed/castor cake @40 kg N/ha; fertilisers @40 kgN/ha; Topping at flower head and harvesting by priming.
- **Cheroot Tobacco:** Spacing of 60 × 45 cm or 60 × 60 cm for transplantation; planting in October middle to November first week; Sheep penning or FYM @10 t/ha; Fertilisers @50:50:100 kg N, P_2O_5 and K_2O/ha (N in 2 equal splits 45 DAP and 60 DAP); Topping at flower head; Harvesting by stalk cut.
- **Cigar Filler:** Spacing of 75 × 50 cm for transplantation; Planting at October middle to November first week; Sheep penning or 25 tons FYM/ha; fertilisers @75:50:100 kg N, P_2O_5 and K_2O/ha (N in 2 equal splits 45 DAP and 60 DAP); Topping at 12–14 leaves; Harvesting by stalk cut.
- **Cigar Wrapper:** Spacing of 90 × 45 cm for transplantation; Planting at October middle to November first week; 15 tons FYM/ha: Fertilisers @125:112:224 kg N, P_2O_5 and K_2O/ha; No topping; Harvesting by priming.
- **Chewing/Bidi Tobacco:** Spacing of 90 × 60 cm for transplantation; Planting at October middle to November first week; FYM @30–40 t/ha; Fertilisers 112: 160: 0 kg N, P_2O_5 and K_2O/ha; topping at 12–14 leaves; harvesting by priming.
- **Hookah and Chewing:** Spacing of 90 × 90 cm for transplantation; Planting in October; Fertilisers @112:112:0 kg N, P_2O_5 and K_2O/ha; topping at 8 leaves; harvesting by priming.

18.12 TERMS USED

- **Priming:** In tobacco, lower leaves mature first and maturity proceeds further up. To ensure that every harvest consists of leaves of fairly uniform maturity, a method of harvest is developed—priming. It consists of harvesting the lower most 2–3–4 leaves at each harvest at weekly interval. The harvesting may be completed in 3–6 harvest, depending how many leaves are retained in each plant.
- **Topping:** Removal of apical growing tip is essential in tobacco to ensure that nutrients and are utilised by specific number of leaves. Depending on the number of leaves left after topping, the leaf quality changes and even nicotine content changes. Hence, topping is recommended at different heights looking to the desirable quality sought in

different tobacco. Generally, lesser the number of leaves—more is the nicotine content and vice versa.

- **Desuckering:** As a result of topping or removal of apical tip, the buds in the axils of branches get activated and grow into branches. These are called **suckers**. Unless their growth is curbed, the photosynthates get wasted and leaf growth is affected. Hence, application of suckericide like Decanal paste is recommended at the site of bud. Most farmers do this by hand removal of bud.
- **Bulking:** It refers to **collecting** the cured tobacco leaves and **stacking** them for a period of 8–10 days, with frequent change over, in such a manner that they do not develop heat or not subjected to fungal attack.
- **Cured leaf yield:** Refers to **tobacco leaf yield per hectare** after curing in specific method.
- **Rabbing:** Process of spreading paddy straw on nursery beds and burning the beds to get rid of pathogens and insects, before manure is applied to the beds.

REFERENCES

Anonymous (2014), *Categories of Tobacco*, http://en.wikipedia.org/wiki/Category:Tobacco_in_India.

Anonymous (2014), *Soil and Climatic Requirements of Tobacco*, Central Tobacco Research Institute, Rajahmundry, http://www.ctri.org.in/pages/ff_soils.html.

Anonymous (2014), *Varieties of Tobacco*, Central Tobacco Research Institute, Rajahmundry, http://www.ctri.org.in/.

Anonymous, 2022, Agricultural statistics at a glance, DAFW, GoI, p. 29.

FAOSTAT, 2023, https://www.fao.org/faostat/en/#data/QCL.

Brockely, M. (2004), Tobacco, *TED Case Studies*, No. 760, http://www1.american.edu/ted/turkish–tobacco.htm.

Jean–Louis Verrier, Aurelia Luciani and Bernard Cailleteau (2007), *Prospects of Orobanche Control in Tobacco*, http://www.imperialtobaccoscience.com/files/pdf/tobaccogeneticstudies/prospects_for_orobanche_control_in_tobacco.pdf.

Chapter 19

Quinoa (*Chenopodium quinoa* L.)

19.1 ECONOMIC IMPORTANCE

Quinoa is an annual herbaceous flowering plant with its grains resembling cereals and leaf appearing like legumes commonly grouped as pseudo-cereal. It is regarded as the food for poor and has a history of about 7000 years of cultivation in the Andean region. Great cultures like the Incas had domesticated and conserved this ancient crop (Jacobsen, 2003). Quinoa was catalogued as one of the most promising crops for humanity during 1996 by FAO considering its properties and multiple uses. It is also considered as an option to solve human nutrition problems (FAO, 2011). The quinoa plant was widely cultivated in the whole Andean region in Columbia, Equator, Peru, Bolivia and Chile before the Spanish conquest. However, the habits and traditional foods of natives were replaced with exotic crops such as wheat and barley. Thereafter, quinoa was cultivated either in small plantations in rural areas for domestic consumption or as border crop in potato or maize. Hence, it was classified as food for poor people.

In order to recognize ancestral practices of Andean communiy, the United Nations General Assembly declared year 2013 as 'The International Year of Quinoa'. Recently, interest has been increased in quinoa crop due to its high-quality protein with a balanced amino acid composition compared to other cereal grains. Quinoa grains are rich source of dietary fiber, vitamins, minerals and natural antioxidants. The protein in quinoa grain is as good in quality as that in cow's milk.

19.2 ORIGIN AND DISTRIBUTION

Quinoa is a potential pseudo-cereal native to the Andean region of South America. It was cultivated and used by the Inca (ruling class) people since 5000 B.C. The major quinoa producing countries are Bolivia, Peru, Ecuador and Chile (Coral and Cusimamani, 2014). In reent years, this crop is cultivated in South America, USA, China, Europe, Canada and India. It is cultivated in the world on an area of 126 thousand hectares with a production of 103 thousand tonnes. Bolivia in South America is the largest producer of quinoa (46%) in the world followed by Peru (42%) and United States of America (6.3%).

In India, quinoa is cultivated in Andhra Pradesh, Uttarakhand and Himalayan region. In Andhra Pradesh, it is successfully grown in Hyderabad and Ananthapuram districts. It is cultivated in an area of 440 hectares with a production of 1053 tonnes. Karnataka and Rajasthana are the potential states emerging for quinoa cultivation in recent times.

19.3 BOTANICAL DESCRIPTION OF PLANT

It is an annual herbaceous plant belongs to Amaranthaceae family. Quinoa is a dicotyledonous annual plant growing to an height of about 1–2 m. It has broad, generally powdery, hairy, lobed leaves, normally arranged alternately. The woody central stem is branched or unbranched depending on the variety and may be green, red or purple. The inflorescence arise from the top or from leaf axils along the stem. Each panicle has a central axis from which a secondary axis emerges either with flowers or bearing a tertiary axis carrying the flowers. These are small, incomplete, sessile flowers of the same colour as the sepals, and both pistillate and perfect forms occur.

Figure 19.1 Quinoa plant.

Figure 19.2 Quinoa inforescence.

19.4 SOIL AND CLIMATIC REQUIREMENTS

Quinoa shows enormous variation and diversity in its adaptation to different environmental conditions and is cultivated up to 4000 M.S.L altitude. It is tolerant to adverse climatic factors such as drought, frost, soil salinity. Its growing season ranges 90-240 days. It grows in the regions with the precipitation ranging from 200-280 mm per year. Though quinoa prefers neutral soils, it can be grown on alkaline soils up to pH of 9.0 and acidic soils up to pH of 5.0. It also thrives in sandy and clay soils. The ideal temperature for quinoa cultivation is around 18°C to 20°C, although it can with stand temperature extremes ranging from 39°C to –8°C. It has greater diversity of adaptation to photo period, altitude, soil pH etc. Quinoa seems to be a quantitative short-day species where in the length of the vegetative period depends not only on the day length and latitude of the origin but also on altitude of the origin (Rishi and Galwey, 1984). The adaptability of quinoa to varied levels of drought is due to the differentiation of a diversity of ecotypes originating in contrasting agro-environments. Plants display various adaptive strategies to drought stress, from morphological to physiological adaptations that serve a range of responses to water deficit, from avoidance to resistance and tolerance. Plants cope up with the drought stress by changing and modifying key physiological processes such as photosynthesis, respiration, water relations, antioxidants and hormone metabolism. Whole-plant responses to drought involve changes in leaf and root growth, in some cases with strong ontogenetic variation. These drought responses at both physiological and morphological levels show intraspecific variation related to ecotype differentiation. Quinoa thus represents an invaluable opportunity, both as a potential crop in consideration of present and future climate change challenges and as an important source of genes with biotechnological applications (Azurita-Silva Andres et al., 2014).

19.5 GENERAL CULTIVATION PRACTICES

Land preparation: The lands will have to be prepared well by 1–2 ploughing with MB ploughs followed by 2–3 harrowing and final levelling. Application of adequate manures is important (2-3 t/ha).

Season and Sowing: Although, quinoa can be cultivated across the seasons, rabi season is ideal. Waterlogging/saturation with heavy rains in kharif poses wilt. Line sowing can be done manually or mechanically at a spacing of 45 cm × 15 cm. Spacing need to be altered based on soil fertility and genotype with a seed requirment ranging from.

Fertilizer application: The nutrient demand of the crop under medium soil fertility is 40:20:20 kg NPK per hectare. However, the dosage can be altered considering the soil fertility and yield potential of the crop.

Inter cultivation and weed control: The crop prefers good aeration. Inter-cultivation and earthing up at 30 DAS is essential. It is always advisable to control the weeds by inter-cultivation.

Pest and disease: Quinoa is resistant to most of the pest and diseases. However, higher moisture/waterlogging during kharif exposes the crop to wilt disease. Draining and drenching with mancozeb can minimize the wilt incidence

Harvesting: Harvesting can be done manually. The crop is harvested by cutting at the base and drying it for 2–3 days. Then they are threshed by beating with poles or mechanically.

REFERENCES

Alvarez-Jubete, L., Arendt, E. and Gallagher, E., 2009, Nutritive value and chemical composition of pseudocereals as gluten free ingredients. *International Journal of food Science and Nutrition*, 60 (1): 240–257

Coral, T. and Cusimamani, E. F., 2014, An andean ancient crop, Chenopodium quiona Willd : A review. Agricultura Tropica Subtropica, 47 (4): 142–146. FAO, 2011, Quinoa: An ancient crop to contribute to world food security, *Technical report of the 37th FAO Conference*, Rome, Italy

Jacobsen, S. E., 2003, The worldwide potential for quinoa (Chenopodium quinoa willd), *Food Reviews International*, 19 (1): 167–177.

Risi, J. and Galway, N. W., 1984, The Chenopodium grains of the Andes: Inca crops for modern agriculture, *Annuals of Applied Biology*, 10 (1): 145–216

MODEL QUESTIONS AND KEY ANSWERS FOR POTATO, TOBACCO AND QUINOA

Choose Most Appropriate Answer and Fill in the Blanks

1. Flue cured tobacco is mainly produced in the state of .... (Odisha/Uttar Pradesh/Maharashtra/Karnataka)
2. Priming is a method of .... (curing/storing/harvesting/planting)
3. Topping tobacco invariably induces growth of .... (leaves/branches/suckers/flowers)
4. Potato dormancy can be broken by the use of .... (urea/glycol ureate/thiourea/uric acid)
5. Most of the potato varieties commercially grown in India are .... potato. (Chilean/Andean/Mexican/Bolivian)
6. The characters of *Nicotiana rustica* include .... leaf. (broad petiolate; narrow petiolate; broad sessile; narrow sessile)
7. Cultivated *Solanum tuberosum* was originated from Solanum .... (*brevicaule/* stomentosum/virdae/andeanum)
8. The poisonous pigment chaconine is found in .... (tobacco leaves/tobacco fruits/potato tubers/potato roots)
9. Pit curing is invariably followed for .... tobacco. (cigarette/cigar/chewing/bidi)
10. Burley tobacco is mainly grown in .... state. (Karnataka/Andhra Pradesh/Gujarat/Tamil Nadu)
11. Maximum potato area is found in .... state. (Uttar Pradesh/Punjab/West Bengal/Madhya Pradesh)
12. .... country has maximum area under tobacco. (Turkey/India/USA/China)
13. VFC tobacco is invariably used to manufacture .... (bidi/cigarette/snuff/gutka)
14. Bidi tobacco is usually .... cured. (sun/air/pit/smoke)
15. Central Tobacco Research Institute is located in .... (Guntur/Hyderabad/Rajahmundry/Warrangal)
16. Cigarette tobacco is invariably grown with .... doses. (low N and low K; low N and high K; high N high K; high N and low K)
17. Bidi tobacco invariably has .... nicotine. (very low/low/medium/high)
18. Bulking is invariably done at .... stage in tobacco. (pre-harvest/pre-curing/post-curing/early growth)
19. Tobacco leaf drying takes place at a temperature of ....°C in flue curing. (30–35/51–60/40–48/68–75)
20. True potato seeds are required at a rate of .... g/ha to establish a commercial crop of potato. (150/750/1500/1850)
21. Lanka tobacco is grown in deltaic islands of .... (Tamil Nadu/Karnataka/Odisha/Andhra Pradesh)
22. Hookah tobacco is mainly grown in the .... state. (West Bengal/Maharashtra/Bihar/Gujarat)

23. Central Potato Research Institute is located in .... (Lucknow/Agra/Nainital/Shimla)
24. Tuberisation in potato is favoured at a temperatures of ....°C. (22–25/17–20/28–30/30–35)
25. .... source of potassium is preferred for VFC tobacco. (MOP/SOP/KNO3/15:15:15)
26. .... disease is more devastating and widespread than other diseases of potato (Early blight/Late blight bacterial wilt/Wart)
27. Cut tuber requirement for planting is in the range of .... q/ha. (8–10/12–15/15–20/25–30)
28. Herbicide recommended for weed control in potato include .... (2,4–D, Atrazine/Alachlor/Diuron)
29. .... is a preferably used source of nitrogen for VFC tobacco (CAN/Urea/Ammonium sulphate/Ammonium chloride).
30. Motihari tobacco grows the state of .... (Uttar Pradesh/Andhra Pradesh/West Bengal/Bihar)
31. Quinoa is a ……………. (Cereal/legume/Pseudo-cereal)
32. UN declared International year of Quinoa during ………… (2005/2013/2025)
33. Quinoa is native to……... (South America/Africa/Asia)
34. Quinoa prefers .... (aeration/water logging/salinity)
35. Wilt in quinoa can be managed with ……….. (Mancozeb/Carboryl/Pendimethalin)

State TRUE or FALSE by Using 'T' or 'F'

1. All varieties of *Nicotiana rustica* are used as chewing or bidi tobacco. (T/F)
2. Potato is a shallow rooted crop. (T/F)
3. Higher application of nitrogen reduces the nicotine content in the tobacco leaves. (T/F)
4. Number of leaves on the tobacco plant decides nicotine content in them. (T/F)
5. Burning quality of tobacco is adversely affected by chlorine. (T/F)
6. Chipsona tubers are rich in sugars. (T/F)
7. Decanol is a popular herbicide used in tobacco. (T/F)
8. Nicotine sulphate in tobacco seeds is used to prepare a type of skin ointment. (T/F)
9. The better aroma and flavour of tobacco leaves is achieved after curing by higher application of calcium and zinc. (T/F)
10. Potato flowers do not bear fertile pollen in short days. (T/F)
11. Potato is planted throughout the year in one or other part of India. (T/F)
12. The tubers are attached to potato plant by stolons. (T/F)
13. Orobenche is an important parasitic weed of tobacco. (T/F)
14. The knots developed on the roots of tobacco are due to bacteria. (T/F)
15. FCV curing of cigarette tobacco causes deforestation. (T/F)
16. Potato tubers contain 80% starch. (T/F)

17. Rabbing is a common technique used to reduce the incidences of disease in main field. (T/F)
18. Priming technique is not adopted in all types of tobacco crops. (T/F)
19. Potato grown in high altitudes generally matures earlier than crop grown in plains. (T/F)
20. Tobacco curing is also called 'colour curing'. (T/F)
21. Quinoa was catalogued as one of the most promising crops for humanity. (T/F)
22. Quinoa is classified as food for rich (T/F)
23. The protein in quinoa grain is as good in quality as that in cow's milk. (T/F)
24. Quinoa belongs to poaceae family (T/F)
25. Quinoa shows enormous variation and plasticity in its adaptation to different environmental conditions. (T/F)
26. The ideal season for quinoa cultivation is summer. (T/F)
27. Ideal NPK ratio for quinoa cultivation is 2:1:1. (T/F)
28. Quinoa is susceptible to pest and diseases. (T/F)
29. Andrapradesh is the major state in India for quinoa production (T/F)
30. Row spacing recommended for quinoa cultivation is 25-30 cm. (T/F)

Answers

Fill in the blanks

1. Karnataka
2. harvesting
3. suckers
4. thiourea
5. Andean
6. broad petiolate
7. brevicaule
8. potato tubers
9. chewing
10. Andhra Pradesh
11. Uttar Pradesh
12. India
13. cigarette
14. sun
15. Rajahmundry
16. low N and high K
17. high
18. post-curing
19. 51–60
20. 150
21. Andhra Pradesh
22. Bihar
23. Shimla
24. 17–20
25. SOP
26. early blight
27. 15–20
28. Alachlor
29. CAN
30. West Bengal
31. Pseudo-cereal
32. 2013
33. South America
34. aeration
35. Mancozeb

True or False

1. F	**2.** T	**3.** F	**4.** T	**5.** T	**6.** F	**7.** F	**8.** T	**9.** F	**10.** T
11. T	**12.** T	**13.** T	**14.** F	**15.** T	**16.** F	**17.** F	**18.** T	**19.** F	**20.** T
21. T	**22.** F	**23.** T	**24.** F	**25.** T	**26.** F	**27.** T	**28.** F	**29.** T	**30.** F

Part 7
FODDER CROPS (FORAGE CROPS)

Chapter 20

Berseem (*Trifolium alexandrium* L.)

20.1 ECONOMIC IMPORTANCE

Berseem is also called by other names **berseem clover** or **Egyptian clover** or **bersim**. It is a **highly nutritive, succulent** winter forage, fed to lactating animals and energising feed for **horses, cattle, camel** and even donkeys. It has 20% crude protein, and 70% dry matter besides being rich in minerals like calcium and phosphorus. Feeding berseem improves the milk yield as well as brings up the weight gain in growing animals. Generally, it is fed to animals along with cereals to improve the nutritive value of latter.

Besides being an important fodder, berseem has other benefits on soil fertility. Its cultivation can improve the nitrogen status, organic carbon content and available phosphorus in the soil. It also loosens a compact soil and helps in better soil aggregation and reduces bulk density, besides improving the soil microbial activity.

It can also **smother** the growth of weeds successfully, and therefore used as a popular sequence crop in severely weed infested plots.

20.2 ORIGIN AND HISTORY

It was originated in **Egypt**. Hence, it is popular as **Egyptian clover**. However, it was domesticated in Syria and later spread to the Mediterranean region and other parts of the world. It appears that it was originated from its progenitors *Trifolium salmoneum* and *Trifolium berytheum,* commonly found in alkaline soils of Nile valley. It was introduced from Egypt to India in 1904. It is a popular winter forage crop in many Asian and African countries.

20.3 AREA AND DISTRIBUTION

Berseem is mainly valued as a **winter crop** in the subtropics as it grows well in **mild winter** and recovers strongly after cutting. It does not grow well under hot summer conditions. It is cultivated from 35°N to the Tropics, from sea level up to 750 m (1500 m in North West Himalaya).

It is an important **winter forage** in subtropical region, but not well suited to temperate and tropical regions. Mostly, it is grown as important forage crop in Mediterranean and Middle

Eastern regions. It is mainly grown in Egypt, Syria, Iran, Pakistan and India, above 35°N latitudes, more successfully. In recent years, area under berseem is increasing fast, as witnessed by its area in Egypt (1.1 million hectare), India (2 million hectare), besides sizable area in Syria, Pakistan, Iran, Morocco and many Southern European countries. Even it is grown successfully in USA, and Australia as an important forage crop. In India, it is regularly grown in the states of Punjab, Delhi, Haryana Rajasthan, Uttar Pradesh and other Western and North Indian states.

20.4 SOIL AND CLIMATIC REQUIREMENTS

It requires **cool dry climate** for its normal growth. It ideally requires 13–15°C for good germination, while its growth is hastened at 18–21°C. Its growth can be reduced at temperature lesser than 15°C and more than 21°C. Frost can make the crop dormant and no regeneration is possible. At temperatures beyond 32°C, the re-growth after cutting is not possible. It cannot withstand either drought or frost. It **cannot be grown in wet or heavy rainfall** regions. It can be grown **in all types of soils** except light sandy soils. It grows better on **well-drained loamy soils** rich in calcium and phosphorus. It can be grown in **alkaline soils,** but not in acid soils. It can be grown up to an altitude of 750 m from the sea level. It can be successfully grown in areas receiving rainfall up to 250 cm.

20.5 BOTANICAL DESCRIPTION OF PLANT

It belongs to family *Fabaceae*. It is a small **shrubby annual** (Figure 20.1) growing to a height of 60–80 cm with shallow root system (45 cm). It has a **hollow and succulent** stem, which can be fibrous after flowering. Leaves are small, tender, slightly pubescent and trifoliate and borne alternatively. The leaflets are small (4–5 cm long and 2–3 cm broad). Flowers are yellowish white and are in elliptical clusters (Figure 20.2) with 2 cm diameter. The flowers are cross-pollinated by honey bees. The fruit is pod, containing single white to purplish red seed.

Figure 20.1 Shrubby crop of berseem.

Figure 20.2 Elliptical clusters of flowers.

It has shallow tap root system (40–50 cm) and its stems are hollow, branching at base. Depending on the branching habits, the cultivars are grouped into two groups. Most popular variety **Miscawi** and **Kahdrawi** belong to branching types.

20.6 VARIETIES

Both **diploid** and **tetraploid** varieties are under cultivation.

Differences between diploid and tetraploid varieties are presented in Table 20.1.

Table 20.1 Difference between Diploid and Tetraploid Varieties

Diploid Varieties	*Tetraploid Varieties*
Same seeds may be used.	Seeds need to be produced every year from breeder.
They cannot withstand severe cold, slow growing, less leafy, but succulent.	They are winter hardy, quick growing, very leafy and succulent.
Re-growth after cutting is possible even during the warm temperatures (> 27°C).	Re-growth after cutting is not possible, if day temperature is > 27°C.
Example: **Miscavi, Chhinchwada, IGFRI 99–1**.	*Example:* **Pusa Giant**.

Important varieties grown in India are:

- **Miscavi:** Most popular diploid variety are grown mostly in the Northern states.
- **Berseem Ludhiana:** Diploid variety, yielding 30% higher than Miscavi.
- **Pupa-Giant:** Tetraploid bred at IARI with dark green broad leaves yielding 10% more than Miscavi.
- **Bundel Berseem-2:** Tolerant to acidity.
- **Vardan:** Diploid, slow growing long duration crop.

In addition, some varieties like BL 10, BL 22, UP Berseem 103, JB 1, JB 2 and JB 3 are also grown.

20.7 CROPPING SYSTEMS

It can be rotated with almost every cereal crop. Some important **rotational systems** are:

Maize–berseem–cow pea: sorghum–berseem–maize: rice–berseem–cow pea cotton–berseem: pearl millet–berseem–maize.

It can also be intercropped with some forage crops like:

Napier grass–berseem: napier grass + berseem + cow pea

20.8 NUTRIENT MANAGEMENT

Like many other legumes, its response to higher level of N application is limited. Hence, a dosage of 20–30 kg N/ha is sufficient. This is supplemented by biological nitrogen fixation, as berseem responds well to the *Rhizobium* treatment. However, application of phosphorus is necessary to the tune of 50–60 kg/ha. More important is addition of 2–3 tons/ha of FYM, which can take care of nutrient needs, besides serving as reason to improve physical characters of soil.

20.9 WATER MANAGEMENT

Invariably, the crop is grown as **irrigated crop** during post-rainy season.

During first month, it is irrigated weekly; it is irrigated at an interval of 12–15 days in next month. Later, it is irrigated weekly after cutting starts. Each irrigation may provide 5 cm water.

20.10 HARVEST AND POST-HARVEST

The crop is suitable for **multi cut growth** and at least 3–4 cuts can be made in a year, when the crop does not experience water stress. First cutting is ready by 60 days and subsequent cutting is possible after 25–35 days after first cut. It is a common practice to leave the crop for seed production after 3–4 cuts. If grains are harvested from the crop without any fodder cut, the crop may mature in 130–160 days. As a fodder crop, it remains on the field for six months.

But, the growth is temperature related. If temperatures are unfavourable, subsequent growth may not be assured. Making hay is widely practiced to store the fodder in compact place for long time. However, it is suitable for silage making (when harvested with low moisture content). Normally, in multiple harvest extending up to a year, green forage yields are around 80 –110 t/ha along with 16–22 tons of dry matter. Seed crop can give 550–600 tons of forage per hectare, in addition to 4–6 q seed/ha.

20.11 PRODUCTION PRACTICES

- **Land preparation:** It needs **fine seed bed**, and hence thorough land preparation by one ploughing with mold board plough or disc plough, clod crushing and two harrowing are necessary.
- **Seeds:** The crop has to be propagated by seeds. The seed requirement varies from 25–30 kg/ha, which may have to be increased up to 35 kg/ha, in case of late or early sowing. As the initial harvest is generally poor, 1–2 kg of fodder sarson (lahi) seeds are mixed with berseem seeds. Similarly, for better growth, both diploid and tetraploid varieties are mixed in the ratio of 1:2. The seeds should invariably treated with *Rhizobium trifolii*.
- **Sowing:** Berseem is sown by two conspicuous methods:
    1. ***Dry bed method*:** Seeds are broadcast on slightly moist soil, mixed, and covered with one cm of fine soil. The seeds germinate and seedlings are established after 10–12 days. Irrigation should be given only after proper germination.
    2. ***Wet bed method*:** Bed of convenient size should be irrigated till field capacity and then the seeds are broadcasted.
- **Season:** First half of October to first week of November.

- **Fertilisers and manures:** 2–3 tons FYM/ha; 20 kg N and 60 kg P_2O_5 per hectare. Full dose applied as basal application.
- **Irrigation:** First irrigation should be given after seedlings emerge and later irrigations should be given once in week for first month. Till 60–70 days, irrigation should be given once in 15 days and later given at every 10 days.

REFERENCES

Anonymous (2009), Berseem, *Feedipedia*, http://www.feedipedia.org/node/248.

Anonymous, 2022, Agricultural statistics at a glance, DAFW, GoI, p. 29.

FAOSTAT, 2023, https://www.fao.org/faostat/en/#data/QCL.

Singh, Chhidda, Prem Singh and Rajbir Singh (2009), *Modern Techniques of Raising Field Crops*, 2nd ed., Oxford and IBH, New Delhi, pp. 437–443.

Chapter 21

Lucerne (*Medicago sativa* L.)

21.1 ECONOMIC IMPORTANCE

It is called **Lucerne** in Europe while in other parts of the world, it is known as **alfalfa**. The Arabic word 'alfalfa' mean **best fodder**. The name 'lucerne' was obtained, when it was being cultivated in a place called **Lucerne of Italy**. It is considered as king of fodders, because of high fodder yields, its high palatability and high nutrients. Relatively, it is resistant to drought and can be grown as irrigated or dry crop. It is grown in areas, where berseem cannot be grown due to less rainfall. It can be grown as an annual or perennial and multi cut or single cut crop. It can supply fodder continuously for 3–4 years. Its green fodder is highly palatable containing 15–20% crude protein and has 72% digestibility of dry weight of harvested fodder. It also contains 1.5% calcium, 0.2% phosphorus and is rich in Vitamins A, B and D. It serves as concentrate and should be advocated in small quantity along with other green fodder. In high doses, it can create **bloat** in cattle. Using Lucerne hay, may even replace the use of grains and concentrates.

21.2 ORIGIN AND HISTORY

It is the oldest cultivated fodder crop, belonging to family *Fabaceae*. It was originated in **Iran**. Hence, it carries Arabic name alfalfa. It must have moved to Greece in 500 BC and later must have spread to other parts of Europe. Spaniards introduced it to USA. It has reached other parts of Asia including India, by 1900.

21.3 AREA AND DISTRIBUTION

It is grown globally in countries like USA, Canada, Argentina, India, Australia, Brazil, New Zealand, Italy and Russia. It can be grown in **cold subtropical** and as well as **tropical** regions. In India, it is grown in the states of Punjab, Haryana, Uttar Pradesh, Gujarat, Maharashtra and Tamil Nadu.

21.4 SOIL AND CLIMATIC REQUIREMENTS

It is highly adaptable to subtropical and tropical regions. Its tolerance limits of temperatures are wide from 15°C to 35°C. It can be successfully grown under rainfed and irrigated situations, but **not in heavy rainfall tracts**, as its tolerance to high humidity is limited. It can be grown at an altitudes up to 2500 m from M.S.L. It can be grown on wide variety of soils (clay to sandy). But, it is successful in well-drained fertile soil, rather than being affected by the texture. It cannot be grown in alkaline soils, but can be grown in slightly acidic soils.

21.5 BOTANICAL DESCRIPTION OF PLANT

It belongs to family *Fabaceae*. It is herbaceous annual or perennial in habit, growing to a height up to 150 cm and has profuse branching habit (up to 40 branches). It has **deep root system** with **deep tap root** and large number of **lateral roots**. The stem is erect and woody at base, while branched arise from crown region. Compound leaves are trifoliate, with middle leaflet having short petiole (Figure 21.1). The leaflets are sharply toothed on upper half. The flowers are purple or blue or yellow or even white (Figure 21.2); they are cross-pollinated (entomophily). Seeds are kidney shaped and light in weight with shiny surface.

Figure 21.1 Trifiliate leaves of Lucerne.

Figure 21.2 Flowers of Lucerne.

21.6 VARIETIES

Important varieties of lucerne adaptable to most lucerne growing states are: Sirsa No. 8, Sirsa No. 9, NDRI selection No. 1 and Rambler. Out of them, Rambler is cold tolerant suitable for hilly regions; Sirsa No. 8 is annual; NDRI selection No. 1 is perennial.

21.7 CROPPING SYSTEMS

It is grown as *rabi* crop in India after maize, sorghum, rice or cowpea + maize or even green gram. Sometimes, it is mixed with oats or intercropped with perennial Napier grass.

21.8 NUTRIENT MANAGEMENT

It is a **heavy feeding forage crop**. Its N requirement is as high as 70 –90 kg/ha. However, a dose of 20–30 kg N is recommended in the form of fertiliser, considering its N fixing abilities through symbiotic fixation. Its phosphorus requirement is as high as 75 kg/ha to facilitate proper nodulation. In perennial varieties for every cut, a dose of nitrogen (15–20 kg/ha) is necessary, while annual dose of phosphorus (75 kg/ha) is required. The crop also requires a heavy dose of manure (25–30 t/ha) for the successful growth.

21.9 WATER MANAGEMENT

In India, it is an **irrigated crop**, as it is principally grown in **post-rainy season**. It requires 15–18 irrigations in year (for perennial crop), most irrigations given during summer. Initial month may need frequent irrigations (once in 7–8 days), while next 2–3 months require irrigation once in 20 days, which may be further shortened to 15 days during summer. When the leaves turn dark, the crop needs irrigation.

21.10 HARVESTING

First cutting is made after 50–60 days after sowing, and subsequent cuts are made after 20–30 days, when crop attains a height of 60 cm. A green biomass yield of 800 –1000 q/ha can be expected. If the crop is grown for seed production, flowering is allowed after January cutting, followed by frequent irrigations till flowering and its withdrawal after the formation of pod.

21.11 PRODUCTION PRACTICES

- **Field preparation:** As it requires **fine seed bed**, a mold board plough followed by two harrowing and final levelling are necessary. Application of manure before final levelling is necessary.
- **Seeds and sowing:** Rhizobium treated seeds should be sown not more than 2 cm deep preferably during September end or latest by the first week of October. Seeds can be broadcasted (using 30 kg seeds/ha) or can be drill sown @ spacing of 20 cm (using (12–15 kg/ha). When soils are heavy (as in Maharashtra) inter row spacing can be as high as 45 cm.
- **Weed management:** As the crop grows slowly till first cutting, a thorough hand weeding (first after 25 days followed by 55 days) is necessary. In case of line sowing, inter-cultivation by hoes is possible.

REFERENCES

Anonymous, 2022, Agricultural statistics at a glance, DAFW, GoI, pp. 29.

FAOSTAT, 2023, https://www.fao.org/faostat/en/#data/QCL.

Kock, G.C. de (2012), Lucerne—King of Fodder Crops, *Grootfontein,* Agricultural Developement Institute, http://gadi.agric.za/articles/Agric/lucerne.php.

Singh, Chhidda, Prem Singh and Rajbir Singh (2009), *Modern Techniques of Raising Field Crops*, 2nd ed., Oxford and IBH, New Delhi, pp. 444–450.

Chapter 22

Oat (*Avena sativa* L.)

22.1 ECONOMIC IMPORTANCE

Oat is an important cereal crop, well known for its fodder/forage value, for its value as healthy human food, for its ability to withstand erosion (as ground cover), as well as for its value as green manure. Oat is used as a popular cattle feed, either as cut fodder or as forage (grazing) or even as hay, besides, its dry straw being consumed by cattle passionately. Its grains are used by human beings in various forms, like rolled/crushed oat meal, flour, oat bread and oat cake; grains are also popularlys fed to horses; oat grains are also brewed as beer in UK and to prepare sweet drink in Latin American countries. Throughout the world, oat is considered as health food–as its grains are rich in dietary soluble fibres (10.6%), proteins (17%), fats (6%), minerals (calcium, iron, manganese, potassium, zinc) and vitamins (B_1, B_2, B_3, B_5 and B_9).

22.2 ORIGIN AND HISTORY

Oat was originated in the **Western Mediterranean region** consisting of Morocco and Spain. But its domestication happened far from origin in Europe some times in Bronze Age. It spread to different temperate countries later. But, hexaploid oat was known to be grown in the foot hills of Himalaya and its reference is found in Mahabharata. But no evidences are found to prove that it originated in India. Nor any evidence is found to show the exact time of introduction of oat to India. The use of oat as feed for horses is mentioned in Ain-i-Akbari published in 1590.

Hexaploid species *Avena sativa* (called white oat) as well as less known *Avena byzantina* oat (called red oat) have common ancestor *Avena strigosa* (diploid).

22.3 AREA AND DISTRIBUTION

Presently, it is spread to **temperate countries** as an annual crop–grown both in spring and autumn. It is grown all over the world, mainly spread in countries like Russia (1.77)million hectare), the USA (0.3 million hectare), Poland (0.49 million hectare), Finland (0.29 million hectare), Kazakistan (0.19 million hectare), Spain (0.46 million hectare) and Romania (0.07 million hectare). It is grown in Southern temperate countries, like Peru, New Zealand, besides Northern temperate countries. It is even grown in many **subtropical** countries like India, Pakistan, Mexico and China (0.2 million hectare). Interestingly, the area under grain oat production has fell from 48 million tons in 1960s to around 25 million tons in 2008. But oat has gained more popularity as fodder crop, ever since its adaptability has widened to subtropical regions.

In the last two decades, oat has gained importance as main fodder crop in many countries of Latin America, Asia and Africa due to twin reasons: its increased adaptability to wider regions other than temperate regions as well as availability of multi cut varieties–suitable for fodder. This phenomenon has resulted in increased area under oat for fodder purposes even in India. Presently, oat is being grown in foot hills of Himalaya in the states of Punjab, Uttar Pradesh, Jammu and Kashmir, mainly for fodder purposes. The scope of growing oat for grains is indeed bright in many subtropical regions and temperate Kashmir valley. But, the productivity of grain oat is not as high as wheat and rice, hence not so popular. Nevertheless, due to its capacity to withstand drought, it may be grown even for grains under dry land situations.

Fodder Oats Production in Jammu & Kashmir State
Oats were introduced to Kashmir valley by Raja Hari Singh in his stud farms during 1940s, but oats became commons farmers' crop in Kashmir only after late seventies, when sub centre of IGFRI and SKUA&T, Srinagar started systematic research. Despite measures to popularise oats in Kashmir during 1970s, including free distribution of seeds, farmers did not take up to oats till early 80s. After large scale demonstrations were carried out throughout the valley during late 80s, many farmers have taken up oats cultivation, on lands which were otherwise kept barren due to extreme cold conditions from October to March.
Now, nearly 25,000 hectare of land is under oat cultivation, mainly for fodder purposes–which has kept fodder supply throughout the year. The Kashmiri farmers have accepted oat crop as an important fodder crop, used as cattle feed—mainly for horses. However, its cultivation for grains is limited, although the temperate conditions of Kashmir valley is ideal to grow oats even for grains.

22.4 SOIL AND CLIMATIC REQUIREMENTS

Oat grows best in cool and humid weather. On an average, it requires a temperature range of 15–24°C during germination, tillering, booting, and heading stages. High temperatures during blooming may promote chaffiness, which obviously reduce the yield. Oat is also known to require more water per unit dry matter among all cereals, except rice. Hence, it is successful in high rainfall regions receiving 50–110 cm rainfall. In the initial slow growth stages, it can even tolerate 1–2 days of frost. But, it is preferred to cultivate it in spring in temperate regions.

It is grown in **all types of soils**, except alkaline and water logged soils, but can tolerate slight acidity. Ideally, oat put up more vegetative growth in loamy soils of course with adequate moisture.

22.5 BOTANICAL DESCRIPTION

Oat belongs to family *Poaceae*, with annual growing habit. In its early phases of growth, the oat may have erect, semi erect or even prostrate growing habits (Figure 22.1), while at later stages, it is erect growing. Like other cereals, tillers emerge out from the basal nodes and form major part of the biomass. The culms (tiller, with a capacity to bear inflorescence independently) are hollow, with lanceolate soft pubescent leaves. Numbers of tillers per plant depends on many growing conditions as well as crop management issues.

Tillers terminate in large inflorescence called **panicle**. Panicle consists of rachis and its branches, each of which ends in spikelet (Figure 22.2), which are subtended by two loose membranes often longer than spikelet itself. The glumes–**lemma** and **palea** protect florets, which are 3 in glumed types of oat and 5–7 in the un-glumed naked types. Two kernels are produced from each spikelet. The oat grain is spindle-shaped caryopsis, covered by soft trichomes and consists of endosperm and germ.

Figure 22.1 Well grown fodder oat crop.

Figure 22.2 The panicles of oat.

22.6 VARIETIES

Being one of the old cereals, grown in many parts of the world, there are large number of varieties—both for grain purposes as well as fodder purposes. In India, oat is more popular as fodder crop than as grain crop. Large number of varieties of oat were developed, since the inception of IGFRI. Some recent fodder oat varieties are:

- **Haryana Javi-8 (HJ-8):** Released for Haryana in 1997; has yield potentiality of 65 t/ha green fodder in two cuts.
- **Sabzar:** Developed by SKUA&T, Srinagar for Kashmir valley and highlands of Jammu; has yield potentiality of 35–40 tons green fodder; can be grown for grains also.
- **Bundel Jai 851:** Variety developed by IGFRI in 1998 as dual purpose variety with profuse tillering more leafy; suitable for four cuts with green fodder yield up to 50 t/ha in 110 days; as grain crop, it takes 140–145 days and has grain yield potentiality up to 1.2 t/ha.
- **Bundel Jai 992:** Variety developed by IGFRI in 2004 for North Western and North Eastern region as single cut variety. It can produce green fodder of 50 t/ha in 105 days or 10 tons of dry fodder after 145 days.

- **Bundel Jai 2004:** Variety released for terrain region of Uttar Pradesh as well as North Western and North Eastern regions as single cut variety in 2002; it has yield potentiality of 50 t/ha of green fodder or 10 t/ha of dry fodder.
- **Harita:** Variety developed by MPKV and released for Maharashtra as multi cut variety, suitable for winter irrigated cultivation; it has green fodder yield of 50 t/ha or 9 t/ha dry fodder.
- **Bundel Jai 991:** Variety developed for hilly zone, as single cut variety by IGFRI, in 2007 with a green fodder yield potentiality of 30 t/ha in 125 days or dry fodder yield of 7 t/ha in 145 days.

22.7 PRODUCTION TECHNIQUES

- **Land preparation:** 2 ploughing + 2 harrowing and levelling
- **Sowing method:** By drill
- **Seed rate:** 80–90 kg/ha spacing 20–25 cm rows
- **Sowing time:** October–November
- **Fertilisers:** 80 kg N: 40 kg P_2O_5/ha
- **Harvesting:** Cut as one time harvest or in multi cut varieties cut, at least 2–3 times (after 80 days and subsequently after 50 days)
- **Yield:** It can yield a green fodder of 35–55 t/ha and grain yield of 1–1.6 t/ha (in case of grain cum fodder varieties).

22.8 OAT QUALITY FOR FODDER

Under Indian conditions, oat is the only crop, which has gained popularity for its grains, green fodder, dry straw, all used as cattle feed, besides oat grains being consumed as human food too.

Grains used as cattle feed

The oat grains may be **naked** (unhulled/hullless) or **hulled**. Both find use as cattle feed, mainly to horses, pigs, and sheep/goat. It is usually powdered and mixed with other concentrate foods. The energy of cattle feed prepared from hullless grains is very high, as compared to hulled grains—obviously due to more concentrated nutrients in endosperm, which get diluted when they are crushed along with hulls. The share of husk (also called **hull**) is as high as 23–30% on weight basis. In some countries, grains with thin husk are preferred for cattle feed.

Oat grains contain 5–7% fat, which no other cereal grains have. This makes oat grain as ideal food for milch animals, fattening animals as well as animals involved in more energy consumption. In addition, its crude protein is also high in oat grains, particularly lysine content. The high crude fibre content of oat is ideal for its use as poultry feed to improve the digestibility. The feeding of oat (to the extent of 20% of total feed) to the egg laying birds can improve the shell of egg and also improve the protein content in eggs.

Oat plant as green/dry fodder

Ideal characteristic for green fodder is high dry matter/higher biomass coupled with higher crude protein and lower crude fibre. The highest dry matter yield is obtained at maturity, by which time, the crude fibre content would have increased and crude protein would have decreased to an undesirable level. For this reason, it is necessary to strike a balance between dry matter and crude fibre/crude protein. It is advisable to cut the single cut fodder variety at 50% flowering or 75–80 days to achieve this balance, to harvest maximum dry matter and higher level of crude protein and low crude fibre. In multi cut varieties, first cut is at 50% flowering and subsequent cuts are made after every 55–60 days—without allowing more crude fibre to accumulate and to allow maximum crude protein.

REFERENCES

Anonymous, FAOSTAT, http://faostat.fao.org/site/567/DesktopDefault.aspx?PageID=567#ancor, 2012.

Anonymous, Health Benefits of Oats, *Whole Grain Council*, http://wholegrainscouncil.org/whole-grains-101/health-benefits-of-oats, 2012.

Anonymous, Indian Rolled Oats-Market Opportunities, Report prepared by Government of Western Australia, pp. 1–7, 2012.

Anonymous, Oat Quality for Animal Fodder, http://www.yara.co.uk/crop–nutrition/crops/oat/crop–nutrition/oat–quality/oat–quality–for–animal–fodder/, 2013.

Anonymous, Oat, Wikipedia, http://en.wikipedia.org/wiki/Oat, 2014.

Anonymous, *Technology for Increasing Forage Production*, IGFRI, Jhansi, pp. 2–7, 1989.

Anonymous, 2022, Agricultural statistics at a glance, DAFW, GoI, pp. 29.

Dost, Muhammad, *Fodder Oats in Pakistan*, FAO, 2012.

FAOSTAT, 2023, https://www.fao.org/faostat/en/#data/QCL.

Kumar, Vinod, Oat Varieties of India, *Agropedia,* http://agropedia.iitk.ac.in/content/oat–varieties–India, 2013.

Misri, Bimal, *Fodder Oats in the Indian Himalaya*, IGFRI, Palampur, pp. 4–14, 2002.

Tiwari, Vinod, Growth and Production of Oat and Rye, Vol. II, Encyclopedia of Life Support Systems, http://www.eolss.net/Eolss–samplAllChapter.aspx, 2010.

MODEL QUESTIONS AND KEY ANSWERS FOR MILLETS

Choose Most Appropriate Answer and Fill in the Blanks

1. Two-row barley is well known for good malting quality, because it contains .... (less sugar/more sugar and more proteins/more sugars and less protein/higher proteins)
2. Wheat grains has .... % of proteins. (5–6/15–20/10–14/3–6)
3. Wheat is well known to contain ...., offering its flour-bread making property. (carotene/gluten/glycol/satarin)
4. Both six-row and two-row barley originated from its progenitor *Hordeum* .... (*sontaneum/santaneum/spontaneum/africana*)
5. Oat was originated in .... regions. (Mexican/Mediterranean/South Asian/North European)
6. Scientific name of six-row barley is *Hordeum* .... (*spontaneum/distyichum/vulgare/indicum*)
7. Historically, oat was regularly fed to .... (buffaloes/cows/horses/sheep)
8. White oat is scientifically called *Avena* .... (*strigosa/spontaneum/dicoccum/sativa*)
9. Most of the grain oat is imported from .... (USA/China/Russia/Australia)
10. Wheat supplies .... of the global food production. (28%/21%/30%/19%)
11. .... is the leading country in the world to produce oat. (USA/Finland/Ireland/Russia)
12. .... state has maximum area and production of barley in India. (Punjab/Haryana/Uttar Pradesh/Bihar)
13. Area under barley cultivation has reduced to one-fourth of the area 50 years back in India, as its area is replaced by .... (chick pea/potato/wheat/oat)
14. Popular einkorn cultivated wheat varieties belong to *Triticum* .... (*dicoccum/monococcum/aestivum/vulgare*)
15. Species of cultivated tetraploid wheat are *Triticum durum* and *Triticum* .... (*monococcum/aestivum/dicoccum/vulgare*)
16. Besides *Triticum aestivum*, other cultivated tetraploid species of wheat cultivated on limited scale is *Triticum* .... (*vulgare/spelta/broccoli/levistia*)
17. Emmer wheat was originated in .... (Europe/South East Iran/South East Iraq/China)
18. Triticale is cross between .... and .... (wheat and barley/barley and rye/wheat and rye/wheat and oat)
19. Triticale was essentially developed with the intention of higher yield and .... (improved quality/resistance to drought/stress resistance/higher fodder yield)
20. More than 50% of wheat production takes place in three states, namely, Punjab, Haryana and .... (Rajasthan/Bihar/Madhya Pradesh/Uttar Pradesh)
21. Spring wheat does not require exposure to ...., while it is essential for winter wheat varieties to promote growth. (high temperatures/long days/cold temperatures/high soil moisture).
22. Bread making wheat is expected to have higher .... content. (protein/gluten/mineral/starch)
23. Higher wheat productivity is witnessed in .... zone of India. (North Eastern/Central/North Western/Northern hilly)

24. Exposure to cold temperatures (0–8°C) for 4-weeks to promote the growth is called .... (cold treatment/cold seasoning/vernalisation/vernalation)
25. Wheat requires .... and .... weather for flowering. (cool and dry/cool and wet/warm and dry/warm and wet)
26. Cloudy weather associated with high humidity can accentuate .... disease in wheat. (blight/rust/smut/leaf spot)
22. The share of hull in oat is as high as .... % on dry basis. (20/25/30/45)
28. The only cereal to have fat content to the tune of 6% is .... (wheat/barley/oat/paddy)
29. .... cereal has shattering problem. (Wheat/Barley/Oat/Paddy)
30. Only cereal that successfully grows and matures at high latitudes up to 70°N and high altitudes (up to 4500 m M.S.L) .... (wheat/barley/oat/paddy)

State TRUE or FALSE by Using 'T' or 'F'

1. In early stages of wheat cultivation, it is difficult to identify weeds like *Phalaris*. (T/F)
2. Barley could be cultivated successfully in any part of India. (T/F)
3. Grains of oat are consumed as cattle feed as well as human food. (T/F)
4. Crown roots are temporary, while seminal roots are permanent in wheat. (T/F)
5. First irrigation is given after 20–21 days of sowing in wheat. (T/F)
6. In the last two decades, oat is gaining more popularity in the whole world as fodder crop. (T/F)
7. Gluten in wheat flour is known for its expanding property. (T/F)
8. Gluten forms around 40% of protein contained in wheat. (T/F)
9. Barley is used to prepare breads. (T/F)
10. Naked oat are not preferred as cattle feed than husked oat. (T/F)

Answers

Fill in the blanks

1. more sugars and less proteins
2. 10–14
3. gluten
4. spontaneum
5. Mediterranean
6. distyichurn/vulgare
7. horses
8. sativa
9. Australia
10. 30%
11. Russia
12. Uttar Pradesh
13. wheat
14. monococcum
15. dicoccum
16. spelta
17. South East Iraq
18. wheat and rye
19. Stress resistance
20. Uttar Pradesh
21. cold temperatures
22. gluten
23. North Western
24. vernalisation
25. warm and dry
26. rust
27. 30%
28. Oat
29. Barley
30. barley

True or False

1. T **2.** F **3.** T **4.** F **5.** T **6.** T **7.** T **8.** F **9.** T **10.** F

Index